LINEAR OPTIMISATION WITH APPLICATIONS

A. M. FITZHARRIS

Grosvenor House
Publishing Limited

This book is published by
Grosvenor House Publishing Ltd
Link House
140 The Broadway, Tolworth, Surrey, KT6 7HT.
www.grosvenorhousepublishing.co.uk

A CIP record for this book
is available from the British Library

ISBN 978-1-78623-571-8

About the Author

Dr Andrew Fitzharris was a Principal Lecturer in Applied and Computational Mathematics at The University of Hertfordshire. Prior to joining the university he was engaged in mathematical and computing work in industry. At one construction company he used linear optimisation to find the optimal mix of the aggregates used in the asphalts laid on a new section of the A1(M) motorway in Hertfordshire and on the runway at RAF Conningsby in Lincolnshire. In addition to his 32-year career teaching on undergraduate and postgraduate courses he also presents mathematics masterclasses for and at The Royal Institution in London.

Dedication

This book is dedicated to John Radcliffe who in 1985 as Head of the Mathematics Group at the then Hatfield Polytechnic, saw my potential as a teacher of mathematics and persuaded the senior management at the Polytechnic to offer me employment. As a result of this I had a long, interesting and rewarding career. I will always be grateful to John for giving me such a wonderful opportunity and for his helpful advice and support during my early years in teaching. Anyone who can find a job they would happily do for nothing is very lucky. I was very lucky and this is due largely to John. He was a wonderful boss and a first-class mathematician.

Preface

Linear optimisation is concerned with the solution of maximisation and minimisation problems that are described by linear functions and linear constraints. In particular, linear programming problems, assignment problems and transportation problems. Problems of this kind arise in areas such as engineering (*e.g.* optimising aggregate blends), economics (*e.g.* linear economic models), financial management (*e.g.* trend channels) and business (*e.g.* product mix, manpower allocation and transport management). This book is intended to introduce this highly applicable branch of applied mathematics to undergraduate and postgraduate students studying mathematics, engineering, operational research and business.

Contents

Part Two - Assignment Problems

11. Solving Assignment Problems

Part Three - Transportation Problems

12. Solving Transportation Problems

Part One - Linear Programming

1. Introduction

___

1.1 Terminology
Consider the following mathematical problem :

$$\text{Maximise :} \quad z = 12x_1 - 4x_2$$
$$\text{Subject to :} \quad 5x_1 + 2x_2 \leq 16$$
$$x_1 \geq 8$$
$$4x_1 + 6x_2 = 12$$
$$x_1, x_2 \geq 0$$

A problem of this kind is called a **linear programming problem** (or **linear programming model** when describing an applied problem). The word **programming** here means **planning** and was used in this context before computer programming was invented. The function $z = 12x_1 - 4x_2$ is called the **objective function**, the inequalities $5x_1 + 2x_2 \leq 16$, $x_1 \geq 8$ and $4x_1 + 6x_2 = 12$ are called the **structural constraints**, the trivial inequalities $x_1, x_2 \geq 0$ are called the **non-negativity conditions** and the variables x_1 and x_2 are called the **structural variables**. The values of x_1 and x_2 that satisfy the structural constraints are called **solutions** of the problem. If these values also satisfy the non-negativity conditions, they are called **feasible solutions** of the problem.

1.2 The General Linear Programming Problem
The general linear programming problem is to optimise (*i.e.* maximise or minimise) an **objective function** in the form :

$$z = c_0 + \sum_{i=1}^{n} c_i x_i, \quad c_i \in \mathbb{R}$$

subject to :

- m **linear constraints**. These can be either :

$$\text{Type 1:} \quad \sum_{j=1}^{n} a_{ij} x_j \leq b_i$$

$$\text{Type 2:} \quad \sum_{j=1}^{n} a_{ij} x_j = b_i$$

$$\text{Type 3:} \quad \sum_{j=1}^{n} a_{ij} x_j \geq b_i$$

where $a_i \in \mathbb{R}$, $b_i \in \mathbb{R}$ and $b_i \geq 0 \;\forall i$. If a b_i value is negative then it can be made positive by multiplying through the inequality by -1.

- **The non-negativity conditions** *i.e.* $x_i \geq 0 \;\forall i$.

A linear programming problem of this kind can be solved using the **graphical method** or a numerical algorithm such as the **simplex method**.

1.3 Historical Background

Problems of this kind were familiar to mathematicians in the 19th century. In 1827 the French mathematician Joseph Fourier published a paper describing an algebraic method for solving linear programming problems. This method is now known as Fourier-Motzkin elimination. In the late 1930's the Soviet mathematician Leonid Kantorovich used linear programming methods for optimising manufacturing schedules and the American economist Wassily Leontief used them for modelling economies. The widespread popularity of these methods began in the early 1940s when they were used for solving the transportation and resource allocation problems experienced by the allies during the second world war. The first numerical algorithm for solving linear programming problems (the simplex method) was developed by the American mathematician George Danzig in 1947. Other significant developments since that time were Duality Theory developed by the Hungarian/American mathematician John von Neumann in 1947/48 and the Interior-Point Algorithm developed by the Indian mathematician Narendra Karmarker in 1984.

1.4 Guidelines for Constructing Linear Programming Models

Unfortunately, no algorithm is available for constructing linear programming models. However, the following guidelines are often useful.

Step 1 : Identify the value to be optimised *i.e.* maximised or minimised and assign a suitable variable.

Step 2 : Determine the values that can be varied in order to produce the optimal solution *i.e.* write down the structural variables.

Step 3 : Write down the objective function.

Step 4 : Identify the factors that limit the value being optimised *i.e.* write down the constraints. These often have the general form :

$$\text{Requirement} \leq \text{Availability}$$

$$\text{Supply} \geq \text{Demand} \quad etc.$$

Step 5 : Write down the restrictions on the values of the structural variables. They can be non-negative, non-positive, unconstrained or integer.

Step 6 : Combine the expressions from the steps above to produce the required linear programming model.

To illustrate the use of these guidelines consider the following manufacturing example.

The Toy Manufacturing Problem

A toy manufacturing company makes bicycles and trucks using three machines *i.e.* a moulder, a lathe and an assembler. The factory manager wishes to calculate the number of bicycles and trucks the company should manufacture each day in order to maximise its daily profit. To do this she is provided with the following information :

- The manufacture of each bicycle requires one hour on the moulder, three hours on the lathe and one hour on the assembler.

- The manufacture of each truck requires one hour on the lathe and one hour on the assembler. The moulder is not used in the manufacture of trucks.

- The moulder is available for three hours each day only.

- The lathe is available for twelve hours each day only.

- The assembler is available for seven hours each day only.

- Everything made at the factory is sold.

- The company makes £8 profit on each bicycle and £5 profit on each truck.

Construct a linear programming model that can be used to determine the optimal solution of this problem.

Solution

Step 1
The value to be optimised is the total daily profit made by the company. The value being optimised in a problem of this kind is often denoted using the variable name z. However, any variable name can be used *e.g. P*.

Step 2
The values that can be varied in order to produce the optimal solution are the number of bicycles and trucks manufactured each day. The structural variables in linear programming problems are often denoted using variable names such as x_1, x_2, *etc*. However, any variable names can be used *e.g. b, t*.

Let x_1 be the number of bicycles manufactured each day and x_2 be the corresponding number of trucks.

Step 3
The objective function is the expression that gives the total daily profit made by the company.

Since the company makes £$8x_1$ profit each day on bicycles and £$5x_2$ profit each day on trucks, the total daily profit z made by the company is :

$$z = 8x_1 + 5x_2$$

Step 4
The factors that limit the value being optimised are the availability of the machines. For example, the lathe is available for twelve hours each day only. Since each bicycle requires three hours on the lathe and each truck requires one hour, the profit made by the company is constrained by the inequality :

$$3x_1 + 1x_2 \leq 12 \quad i.e. \quad 3x_1 + x_2 \leq 12$$

Call this the lathe constraint. By the same argument, the moulder constraint is :

$$1x_1 + 0x_2 \leq 3 \quad i.e. \quad x_1 \leq 3$$

and the assembler constraint is :

$$1x_1 + 1x_2 \leq 7 \quad i.e. \quad x_1 + x_2 \leq 7$$

3

Step 5

Since the company cannot manufacture a negative number of toys the restrictions on the values of the structural variables are :

$$x_1 \geq 0 \quad \text{and} \quad x_2 \geq 0$$

Step 6

Combining the expressions from the steps above the linear programming model of the toy manufacturing problem is :

$$\text{Maximise :} \quad z = 8x_1 + 5x_2$$
$$\text{Subject to :} \quad 3x_1 + x_2 \leq 12$$
$$x_1 \leq 3$$
$$x_1 + x_2 \leq 7$$
$$x_1, x_2 \geq 0$$

1.5 The Graphical Method

The graphical method can be used for solving linear programming problems that contain **two** structural variables *e.g.* x_1, x_2.

Drawing Straight Lines

The graphical method involves drawing the graphs of linear equations such as $4x_1 + 8x_2 = 10$. To illustrate the procedure used to do this it will be assumed that x_1 is plotted **horizontally** and x_2 is plotted **vertically**.

Firstly, find the coordinates of two points that lie on the straight line. The easiest way to do this is to let $x_1 = 0$ and solve for x_2 and then to let $x_2 = 0$ and solve for x_1. For the equation above :

when $x_1 = 0$, $8x_2 = 10$, $x_2 = 10/8$ *i.e.* $x_2 = 5/4$ **and** when $x_2 = 0$, $4x_1 = 10$, $x_1 = 10/4$ *i.e.* $x_1 = 5/2$

Finally, mark the coordinates of the two points (*e.g.* $(0, 5/4)$ and $(5/2, 0)$) on the graph and join them with a straight line *e.g.*

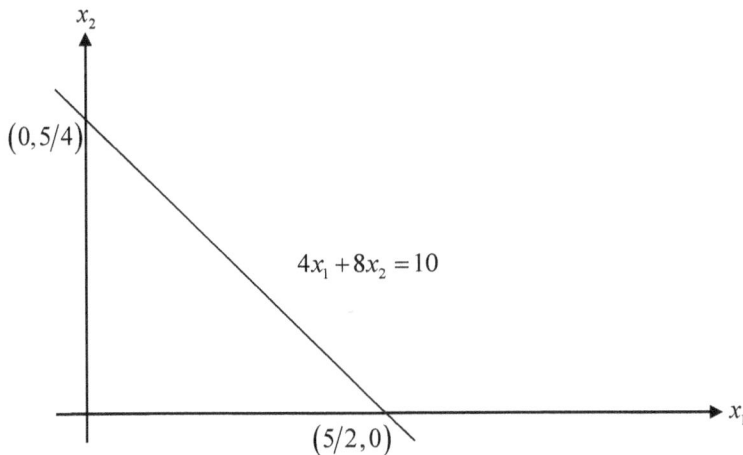

4

Note : When drawing straight lines always use a **ruler** and a **sharp pencil** !

Supplementary Exercise
Draw the graph of the equation $4x_1 + 2x_2 = 16$. Plot x_1 horizontally and x_2 vertically. Choose scales for the axes so that the graph fills as much of the page as possible.

Solution
The graph passes through the points $(0,8)$ and $(4,0)$.

The Procedure
The graphical method is based upon the fact that the solutions of a linear programming problem (*i.e.* the values of x_1 and x_2 that satisfy the constraints) form an area in the $x_1 - x_2$ plane. This area is called the **feasible region**. Since the inequalities in a linear programming problem are always **weak** (*i.e.* $\leq$ or $\geq$), the values of x_1 and x_2 that lie on the **boundary** of this area are also part of the feasible region.

To sketch the feasible region for the toy manufacturing problem consider each of the constraints in turn.

- The non-negativity conditions *i.e.* $x_1 \geq 0$ and $x_2 \geq 0$ show that the feasible region must lie in the first quadrant of the $x_1 - x_2$ plane.

- The graph of $3x_1 + x_2 = 12$ has the form :

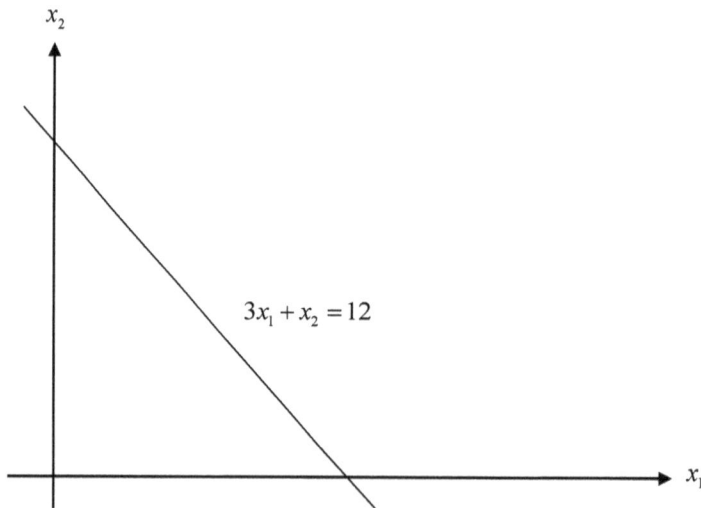

The values of x_1 and x_2 that satisfy the constraint $3x_1 + x_2 \leq 12$ lie on <u>and</u> to one side of this line. To determine <u>which side</u> of the line take a general point and see if its coordinates satisfy the inequality. The easiest point to take in this problem is the origin *i.e.* the point $x_1 = 0$, $x_2 = 0$. Substituting these values into the left-hand side of the inequality gives $0 + 0$ *i.e.* 0. Clearly this value is ≤ 12 and hence the values of x_1 and x_2 that satisfy this constraint lie inside and on the boundary of the closed area formed by the line $3x_1 + x_2 = 12$ and the axes.

- The graph of $x_1 = 3$ has the form :

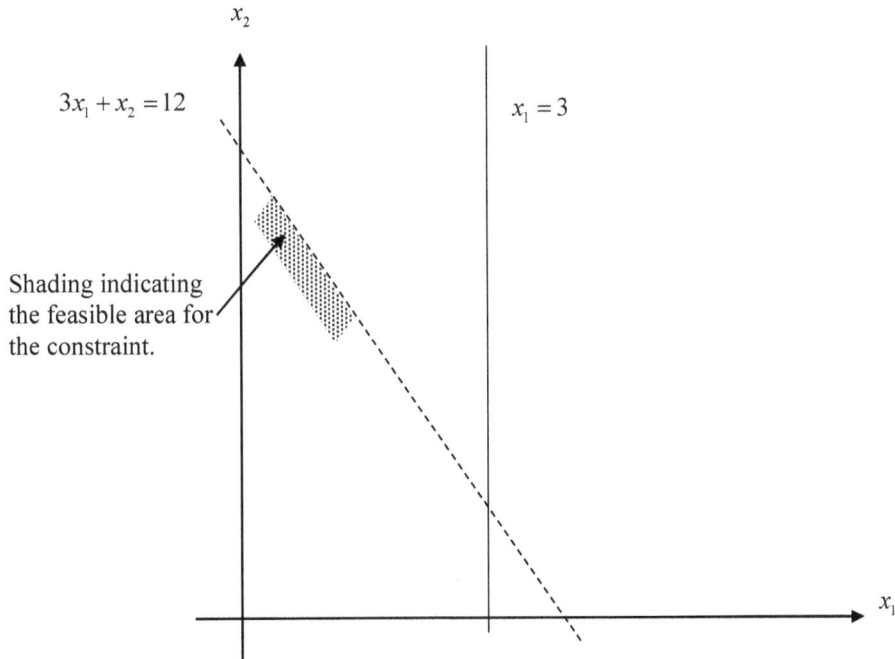

$3x_1 + x_2 = 12$

$x_1 = 3$

x_2

Shading indicating
the feasible area for
the constraint.

x_1

By the same argument the values of x_1 and x_2 that satisfy the constraint $x_1 \leq 3$ must lie on
and inside the boundary of the open area formed by the line $x_1 = 3$ and the axes.

- The graph of $x_1 + x_2 = 7$ has the form :

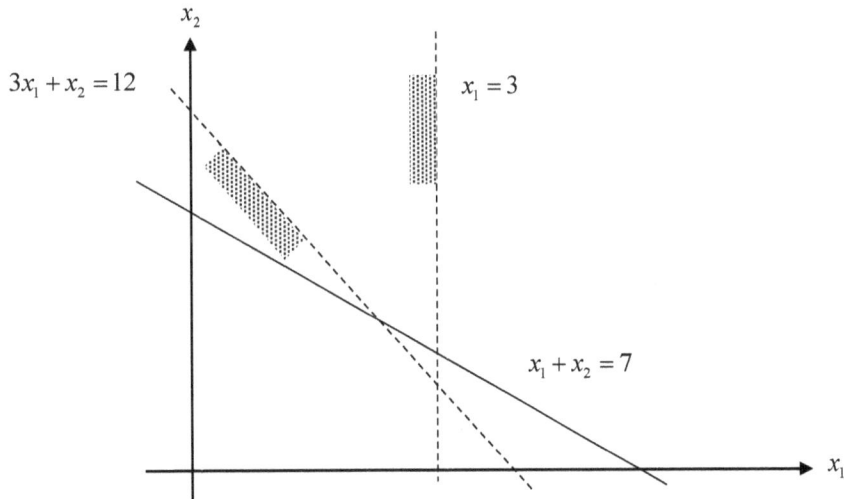

$3x_1 + x_2 = 12$

$x_1 = 3$

x_2

$x_1 + x_2 = 7$

x_1

By the same argument the values of x_1 and x_2 that satisfy the constraint $x_1 + x_2 \leq 7$ must lie on
and inside the boundary of the closed area formed by the line $x_1 + x_2 = 7$ and the axes.

In a linear programming problem **all** of the constraints must be satisfied simultaneously. Hence, the feasible region is the area that is **common** to all of those identified so far. Hence, the feasible region for the toy manufacturing problem is :

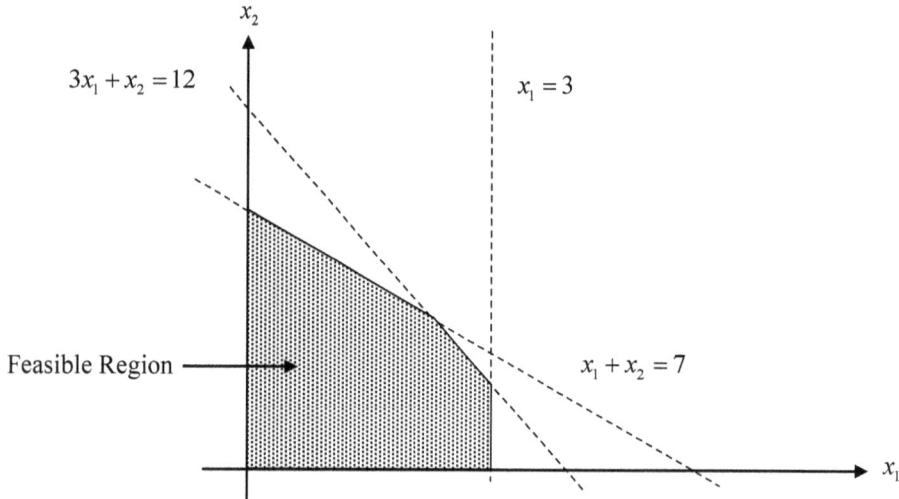

To complete the solution the point inside <u>or</u> on the boundary of the feasible region that gives the maximum value of z (*i.e.* the maximum daily profit) must be found. To do this choose an arbitrary value of z *e.g.* $z = 10$ and see if there are any feasible points that give this value.

Adding the line $8x_1 + 5x_2 = 10$ to the graph it can be seen that it does intersect the feasible region *i.e.*

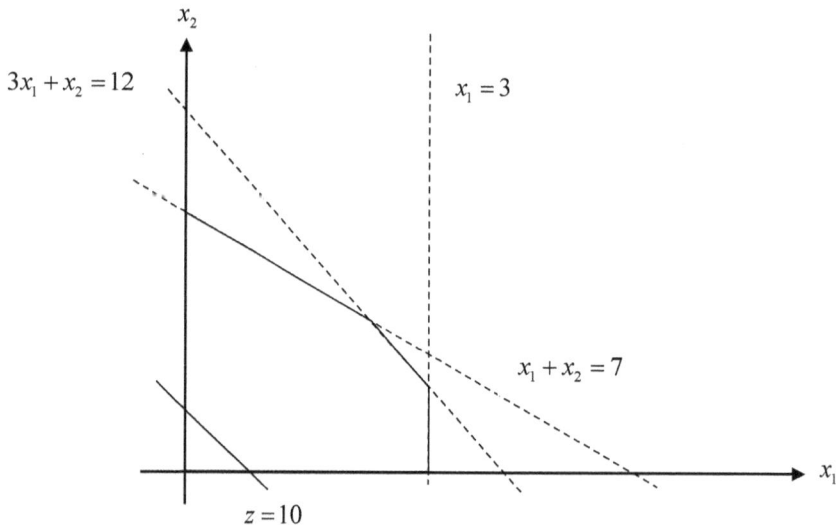

Hence, there are feasible points for which $z = 10$ *i.e.* the maximum daily profit is **at least** £10. This operation can now be repeated with larger values of z. By adding the lines $8x_1 + 5x_2 = 20$,

$8x_1 + 5x_2 = 30$ and $8x_1 + 5x_2 = 40$ to the graph it can be seen that there are feasible points for which $z = 20$, $z = 30$ and $z = 40$ *i.e.*

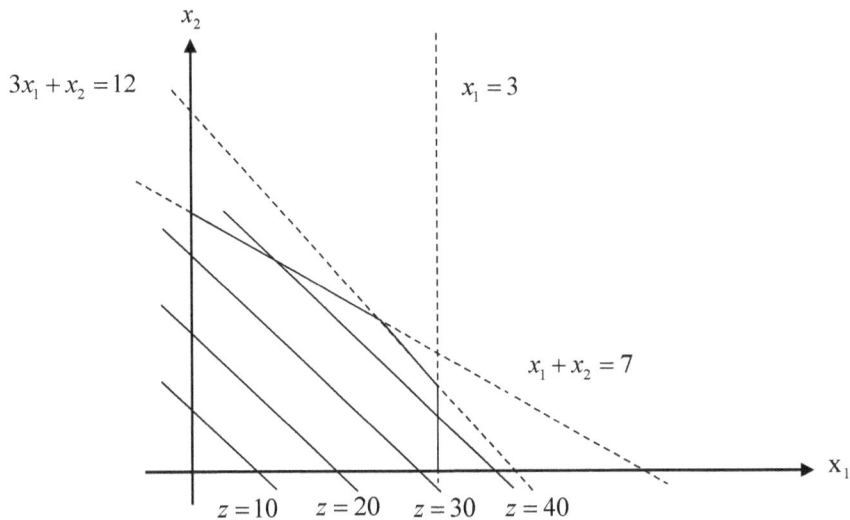

Hence, the maximum daily profit is **at least** £40.

By looking at the graph above it can be seen that as the value of z is **increases** the objective function moves **away** from the origin. Hence, to find the maximum daily profit this line must be moved as far from the origin as possible while ensuring that it remains inside <u>or</u> on the boundary of the feasible region.

As the objective function is moved away from the origin it can be seen that the furthest feasible point lies at the intersection of the lines $3x_1 + x_2 = 12$ and $x_1 + x_2 = 7$ *i.e.*

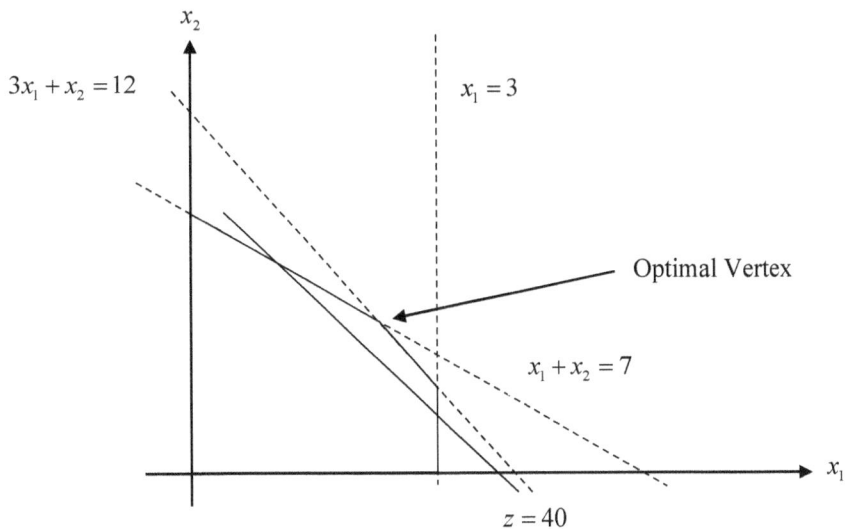

This point is called the **optimal vertex**. It is a feature of all non-integer constrained linear programming problems that the optimal solution occurs on the boundary of the feasible region and usually occurs at a vertex. This fact helps considerably when solving larger and more complicated linear programming problems. To find the optimal solution the coordinates of the optimal vertex can be read off from the graph. However, it is more accurate to find the coordinates by solving simultaneously the equations that intersect at the optimal vertex. In this problem these are the equations :

$$3x_1 + x_2 = 12$$
$$x_1 + x_2 = 7$$

The solutions of these equations are $x_1 = 2.5$ and $x_2 = 4.5$. Substituting these values into the objective function the corresponding value of z is 42.5. Hence, to maximise its daily profit the toy manufacturing company should produce 5 bicycles and 9 trucks every 2 days. Its maximum daily profit is then £42.50.

Notes
- In a minimisation problem the value of z must be **decreased**.

- In practice it is only necessary to add **two** z-lines to the diagram *i.e.* one to establish the gradient of the objective function and the other to establish the direction of movement as the value of z is increased or decreased. Once this information has been established the optimal vertex can be found by moving either line as far as possible in the appropriate direction while ensuring that it remains inside <u>or</u> on the boundary of the feasible region.

- The solution of integer constrained linear programming problems is discussed in Chapter 8.

1.6 Problem Cases
In the toy manufacturing problem the feasible region was a closed region in the $x_1 - x_2$ plane and the problem had a unique solution. However, other types of linear programming problem can arise.

Non-Unique Solutions
This situation arises when the objective function is parallel to one side of the feasible region. For example, consider the following problem :

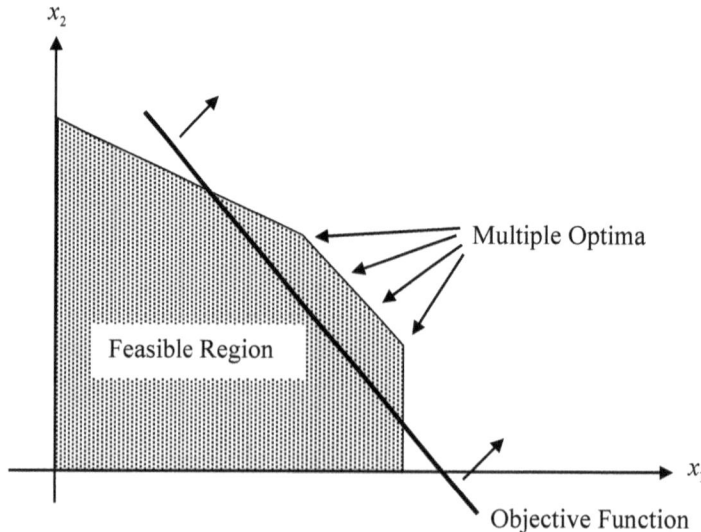

Here all of the x_1 and x_2 values along the parallel side give the optimal value of z and hence this problem has infinitely many solutions.

Infeasible Problems

A linear programming problem is said to be **infeasible** when the feasible region is empty. This situation arises when the constraints are **mutually contradictory** *i.e.* contradict each other. Unfortunately, this situation is not always easy to identify. For example, it is not immediately obvious that the following constraints :

$$x_1 \leq 2$$
$$-x_1 + x_2 \leq 1$$
$$x_1 + x_2 \geq 8$$

contradict each other. However, sketching them and identifying their feasible values *i.e.*

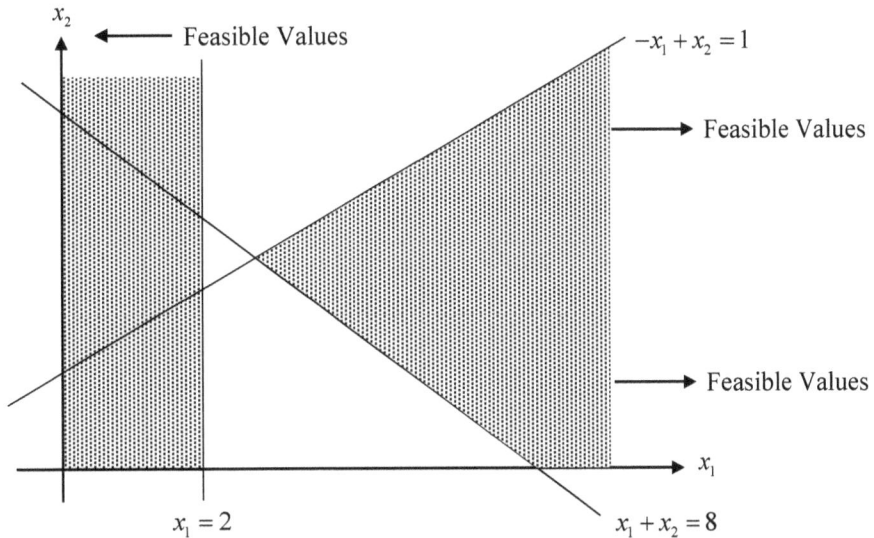

it can be seen that there is no region in the $x_1 - x_2$ plane in which they are all satisfied simultaneously. Hence, the feasible region is empty *i.e.* the associated linear programming problem is infeasible.

The Feasible Region is Open-Ended
In problems of this kind there are <u>three</u> possibilities *i.e.*

Case 1 : The problem can have a unique solution *e.g.*

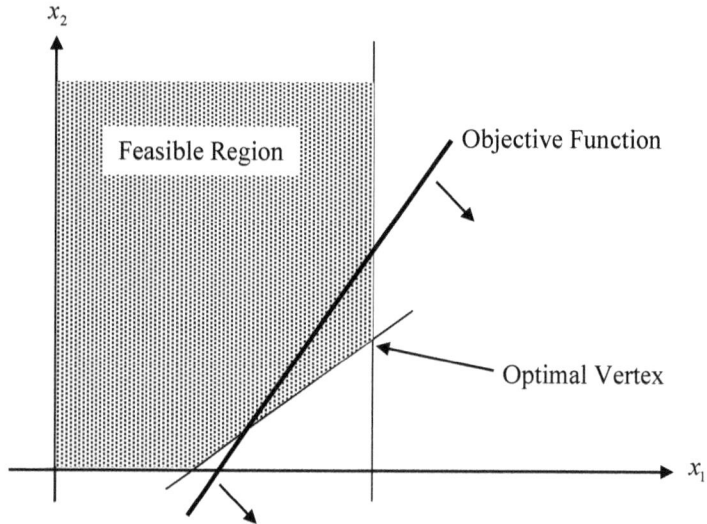

Case 2 : The problem can have infinitely many solutions *e.g.*

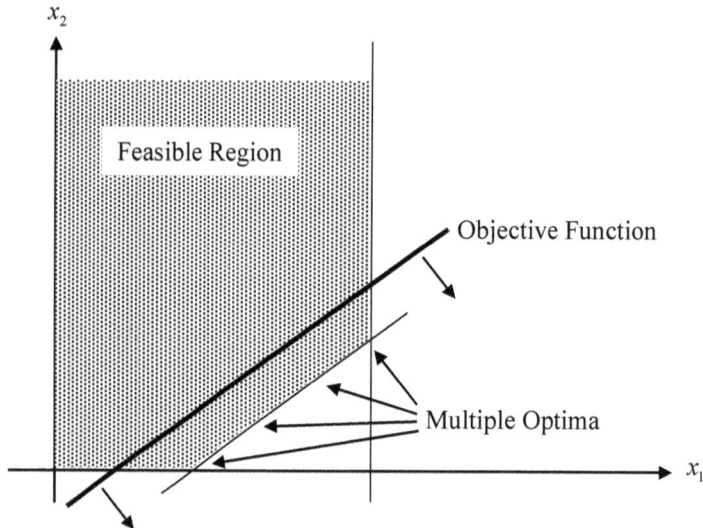

Case 3 : The problem can be unbounded *e.g.*

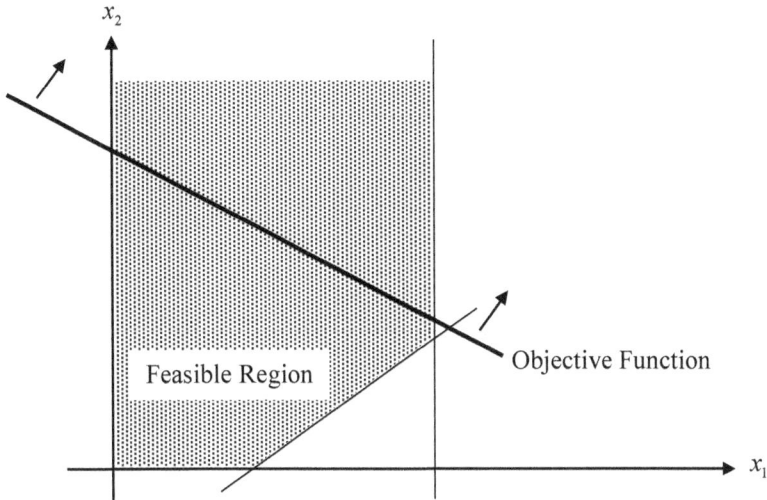

In this case the constraints will still be satisfied even if objective function becomes infinitely large (or small).

In Chapter 3 it will be shown how these problem cases can be recognised when solving linear programming problems using numerical algorithms such as the simplex method.

1.7 Exercises 1a

1. Use the graphical method to solve the following linear programming problems :

    (i)   Maximise $z = 2x_1 + x_2$
           Subject to : $x_2 \leq 5$
                         $x_1 + 2x_2 \leq 12$
                         $x_1, x_2 \geq 0$

    (ii)   Maximise $z = 2x_1 + 3x_2$
           Subject to : $2x_1 + x_2 \leq 8$
                         $x_1 + x_2 \leq 6$
                         $x_1 + 2x_2 \leq 10$
                         $x_1, x_2 \geq 0$

    (iii) Minimise $z = 2x_1 + 3x_2$
           Subject to : $4x_1 + x_2 \geq 6$
                         $x_1 + 2x_2 \geq 5$
                         $x_1 + 5x_2 \geq 8$
                         $x_1, x_2 \geq 0$

2. A small factory manufactures two types of cloth called Standard and Deluxe which it sells with profits of £1.00 and £1.50 per kg respectively. Standard is manufactured from grey wool, red wool and green wool in the proportions 0.75 to 0.125 to 0.125. Deluxe is manufactured from the same wools in the proportions 0.5 to 0.333 to 0.167. Each week the manager can buy up to 750kg of grey wool, 200kg of red wool and 130kg of green wool. To maximise the weekly profit made by the company the manager uses the following linear programming model :

$$\begin{aligned} \text{Maximise} \quad & z = x_1 + 1.5x_2 \\ \text{Subject to :} \quad & 0.75x_1 + 0.5x_2 \le 750 \\ & 0.125x_1 + 0.333x_2 \le 200 \\ & 0.125x_1 + 0.167x_2 \le 130 \\ & x_1, x_2 \ge 0 \end{aligned}$$

(i)   What do the variables x_1 and x_2 represent ?

(ii) Describe the meaning of each expression in the model.

(iii) The optimal solution of the linear programming model is $x_1 = 480$, $x_2 = 420$ and $z = 1110$. Use this solution to determine the strategy that should be adopted by the manager.

3. Jamesons Electrics Ltd. employs two part-time service engineers Robyn and Laura, to repair faulty televisions, DVD players and radios. At the beginning of each week the manager decides how many days Robyn and Laura should be employed during that week. Robyn is paid £25 per day and can repair 1 television, 3 DVD players and 3 radios each day. Laura is paid £22 per day and can repair 1 television, 2 DVD players and 6 radios each day. At the beginning of a particular week there are 5 televisions, 12 DVD players and 18 radios awaiting repair.

(i)   Formulate a linear programming model from which the manager can determine how he should employ Robyn and Laura in order to minimise the cost of their wages.

(ii) Use the graphical method to solve your linear programming model from 3(i) and interpret the solution you obtain.

4. The Dreams bedding company manufactures two high-quality continental quilts, the Majestic and the Royal using two machines, called the Maker and the Embroiderer. The Majestic requires 4 minutes on the Maker and 3 minutes on the Embroiderer. The Royal requires 2 minutes on the Maker and 5 minutes on the Embroiderer. The company makes £12 profit on each Majestic and £10 profit on each Royal. The Maker is available for 2000 minutes each week and the Embroiderer is available for 3000 minutes each week. The company has back orders for 100 Majestics and 200 Royals that must be satisfied.

(i)   Formulate a linear programming model from which the manager can determine the number of each type of quilt the company should manufacture each week in order to maximise its total profit.

(ii) Use the graphical method to solve your linear programming model from 4(i) and interpret the solution you obtain.

(iii) If the profit made on each Royal is fixed at £10, over what range can the profit made on each Majestic vary if the solution obtained in 4(ii) is to remain optimal ?

(iv) If the Maker becomes available for 2200 minutes each week (at no extra cost), what is the new optimal product mix ?

(v) If the increased capacity is available only if the company rents an additional Maker, what is the maximum rent the company would be prepared to pay ?

5. Use the graphical method to solve the following linear programming problems :

(i) Maximise $z = -8x_1 + 4x_2$
    Subject to : $x_1 - x_2 \leq 2$
                  $2x_1 - x_2 \geq -3$
                  $x_1 - x_2 \geq -4$
                  $x_1, x_2 \geq 0$

(ii) Maximise $z = x_1 + x_2$
    Subject to : $x_2 \geq 2$
                  $x_1 \leq 2$
                  $x_1 - x_2 \geq 1$
                  $x_1, x_2 \geq 0$

(iii) Maximise $z = x_1 + x_2$
    Subject to : $2x_1 - x_2 \geq -1$
                  $x_1 - 2x_2 \leq 2$
                  $x_1, x_2 \geq 0$

1.8 Numerical Methods of Solution

Most linear programming problems are solved using iterative numerical algorithms. To begin the study of these methods problems that contain **type 1** (*i.e.* $\leq$) constraints only will be considered. The procedures for dealing with problems that may also contain **type 2** (*i.e.* $=$) constraints and **type 3** (*i.e.* $\geq$) constraints will be considered in Chapter 2.

Canonical Form

To be able to use numerical algorithms for solving linear programming problems all problems of this kind must be converted to a standard format called **canonical form**. To do this :

- Convert the problem into a minimisation problem if necessary. If the objective function is to be maximised then multiply it by -1 and minimise the function that results *e.g.* to maximise :

$$z = f(x_1, x_2, \ldots, x_n) + \text{constant}$$

minimise :

$$\overline{z} = -f(x_1, x_2, \ldots, x_n) - \text{constant}$$

The maximum value of z is then the minimum value of $-\overline{z}$.

- Convert the objective function into an equation *i.e.* transpose it into the form :

$$-z + f(x_1, x_2, \ldots, x_n) = \text{- constant}$$

or :

$$-\bar{z} - f(x_1, x_2, \ldots, x_n) = \text{constant}$$

i.e. take the *z*-term over to join the *x*-terms and take the constant term over to the other side. Adopting this convention allows the solution of a maximisation problem to be read off directly (*i.e.* without the need for reconversion).

- Convert each **type 1** constraint into an equation by adding a non-negative **slack variable**. If the constraint represents the limit on the availability of a resource then the slack variable will represent the unused amount of that resource.

- Write down the non-negativity conditions. These will now include all of the slack variables added during the conversion.

Example 1.1
Write the following linear programming problem in canonical form :

$$
\begin{aligned}
\text{Maximise} \quad & z = 7 - x_1 - x_2 + 2x_3 - x_4 - 3x_5 \\
\text{Subject to :} \quad & x_1 + x_2 + x_3 + x_4 + x_5 \leq 15 \qquad \text{---- (1)} \\
& x_1 - x_2 - 2x_3 + 2x_4 \leq 3 \qquad \text{---- (2)} \\
& 2x_1 + 3x_2 - x_3 - x_4 - x_5 \leq 2 \qquad \text{---- (3)} \\
& -x_1 - x_2 + 3x_3 + x_4 + 2x_5 \geq -14 \qquad \text{---- (4)} \\
& x_1, x_2, x_3, x_4, x_5 \geq 0
\end{aligned}
$$

Solution
Multiplying through the objective function by -1 to convert the problem into a minimisation problem :

$$\bar{z} = -7 + x_1 + x_2 - 2x_3 + x_4 + 3x_5$$

Transposing :

$$-\bar{z} + x_1 + x_2 - 2x_3 + x_4 + 3x_5 = 7$$

Adding a slack variable x_6 to constraint (1) :

$$x_1 + x_2 + x_3 + x_4 + x_5 + x_6 = 15$$

Adding a slack variable x_7 to constraint (2) :

$$x_1 - x_2 - 2x_3 + 2x_4 + x_7 = 3$$

Adding a slack variable x_8 to constraint (3) :

$$2x_1 + 3x_2 - x_3 - x_4 - x_5 + x_8 = 2$$

Multiplying through constraint (4) by -1 (to make the right-hand side positive) and adding a slack variable x_9 :

$$x_1 + x_2 - 3x_3 - x_4 - 2x_5 + x_9 = 14$$

The non-negativity conditions are :

$$x_1, x_2, x_3, x_4, x_5, x_6, x_7, x_8, x_9 \geq 0$$

Combining these, the canonical form of the linear programming problem becomes :

Minimise $-\bar{z} + x_1 + x_2 - 2x_3 + x_4 + 3x_5 = 7$

Subject to : $x_1 + x_2 + x_3 + x_4 + x_5 + x_6 = 15$

$$x_1 - x_2 - 2x_3 + 2x_4 + x_7 = 3$$

$$2x_1 + 3x_2 - x_3 - x_4 - x_5 + x_8 = 2$$

$$x_1 + x_2 - 3x_3 - x_4 - 2x_5 + x_9 = 14$$

$$x_1, x_2, x_3, x_4, x_5, x_6, x_7, x_8, x_9 \geq 0$$

Note

Once a linear programming problem has been written in canonical form there will always be fewer equations (*i.e.* constraints) than unknowns. For example, in the problem above there are **four** equations in **nine** unknowns. Hence, there are (9 - 4) *i.e.* five **degrees of freedom.** This means that the linear programming problem can be solved by assigning arbitrary values to five of the variables and then solving the resulting equations for the other four. The arbitrary value used is usually **zero.** A solution obtained in this way is not always the optimal solution of the problem.

1.9 Further Terminology

- A solution of a linear programming problem that is obtained by setting some of the variables to zero and then solving for the others is called a **basic solution**.

- A basic solution that satisfies the non-negativity conditions is called a **basic feasible solution**.

- The variables whose value is set to zero are called **non-basic** variables.

- The other variables *i.e.* the variables for which the system is solved are called **basic** variables. The set containing the basic variables in a linear programming problem is called the **basis**.

- This terminology can be used to give a formal definition of the term **canonical form** :

  A linear programming problem is said to be in **canonical form** if it is written as a minimisation problem in which each constraint is an equation that contains a basic variable with a coefficient of one.

  The linear programming problem above satisfies this definition since the slack variables x_6, x_7, x_8 and x_9 can be treated as basic variables.

Example 1.2
Consider the following linear programming problem (in canonical form) :

$$\text{Minimise} \quad -z + 3x_1 + 4x_2 = 0$$
$$\text{Subject to :} \quad x_1 + x_2 + x_3 = 5$$
$$2x_1 - x_2 + x_4 = 7$$
$$x_1, x_2, x_3, x_4 \geq 0$$

Here, there are two equations in four unknowns *i.e.* there are two degrees of freedom. If the structural variables x_1 and x_2 are treated as non-basic variables and the slack variables x_3 and x_4 are treated as basic variables then a basic solution of this problem is :

$$x_1 = x_2 = 0, \ x_3 = 5 \text{ and } x_4 = 7$$

Since this solution satisfies the non-negativity conditions it is also a basic feasible solution.

1.10 The Simplex Method
In the description of the graphical method for solving linear programming problems in Chapter 1 it was seen that for a non-integer constrained problem, the optimal solution always lies on the edge of the feasible region and usually occurs at a vertex. The simplex method is based on this idea. The method starts at an initial **basic feasible solution** (bfs) and then moves systematically from one feasible corner to another until the optimal solution is found. When the linear programming problem contains type 1 (*i.e.* $\leq$) constraints only the **origin** is always the initial basic feasible solution. Hence, for a two-variable problem the simplex method will converge to the optimal solution in the following way (although the actual route taken may vary) :

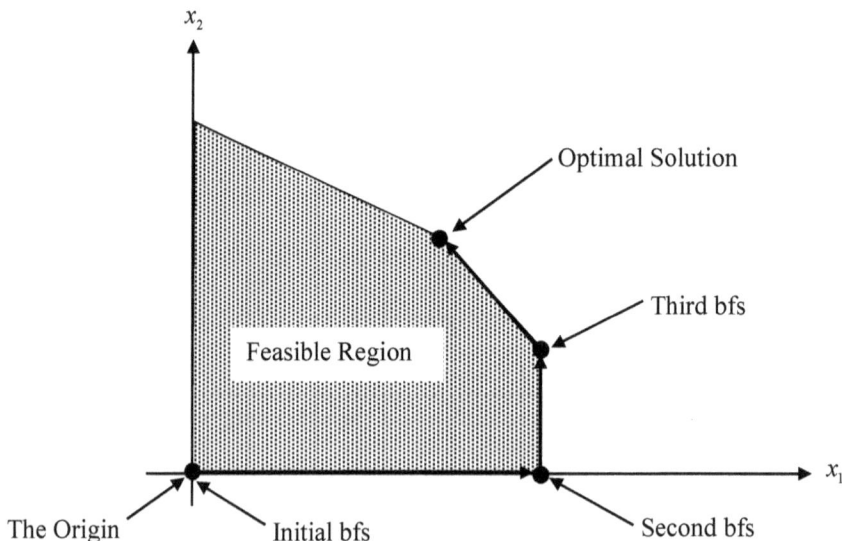

Note
An alternative way of solving a linear programming problem is to calculate the coordinates of the vertices of the feasible region and then to compare the values of the objective function at each one. However, this is an extremely **inefficient** method of solution and the simplex method will always find the optimal solution with less work.

Development of the Algorithm

To develop the algorithm used in the simplex method consider the following two-variable problem :

$$
\begin{aligned}
\text{Maximise} \quad & z = 5x_1 + 9x_2 \\
\text{Subject to :} \quad & 3x_1 + 4x_2 \leq 2400 \\
& x_1 + 2x_2 \leq 900 \\
& 2x_1 + 3x_2 \leq 1600 \\
& 2x_1 \leq 1200 \\
& x_1, x_2 \geq 0
\end{aligned}
$$

This problem can be written in canonical form as :

$$
\begin{aligned}
\text{Minimise} \quad & z - 5x_1 - 9x_2 = 0 \\
\text{Subject to :} \quad & 3x_1 + 4x_2 + x_3 = 2400 \qquad \text{---- (1)} \\
& x_1 + 2x_2 + x_4 = 900 \qquad \text{---- (2)} \\
& 2x_1 + 3x_2 + x_5 = 1600 \qquad \text{---- (3)} \\
& 2x_1 + x_6 = 1200 \qquad \text{---- (4)} \\
& x_1, x_2, x_3, x_4, x_5, x_6 \geq 0
\end{aligned}
$$

To aid clarity the $\bar{z}$ notation will not be used.

To obtain the initial basic feasible solution let the **structural** variables be **non-basic** and the **slack** variables be **basic**. This gives the initial solution :

$$
x_1 = x_2 = 0, \; x_3 = 2400, \; x_4 = 900, \; x_5 = 1600, \; x_6 = 1200 \text{ and } z = 0
$$

However, this is **not** the optimal solution of the problem. In the objective function the coefficients of x_1 and x_2 are negative *i.e.* increasing x_1 and x_2 will reduce the value of z further.

The algebraic equivalent of moving from one feasible corner to another is to increase <u>**one**</u> of the non-basic variables from zero to its largest possible value. The objective function is :

$$
z - 5x_1 - 9x_2 = 0
$$

To reduce z by the largest amount x_2 must be increased (since this has the largest negative coefficient). The value of x_1 will remain zero.

Key Point

*The variable that reduces z by the largest amount is the variable that has the **largest negative** coefficient in the objective function.*

This variable is called the **pivot variable**. The column containing the pivot variable is called the **pivot column**.

The new values of the variables must satisfy **all** of the constraints in the problem. The options for increasing x_2 are therefore :

a. (1) $\Rightarrow$ increase x_2 to $2400 \div 4$ *i.e.* to 600 and reduce x_3 to 0

b. (2) $\Rightarrow$ increase x_2 to $900 \div 2$ *i.e.* to 450 and reduce x_4 to 0

c. (3) $\Rightarrow$ increase x_2 to $1600 \div 3$ *i.e.* to 533.3 and reduce x_5 to 0

Note : (4) $\Rightarrow$ **nothing** about the increase in x_2

To ensure that the new values of the variables satisfy the non-negativity conditions option **b** must be chosen. Increasing x_2 beyond 450 will make one or more of the basic variables negative (*e.g.* x_4).

Key Point
*To find the **largest** increase in the pivot variable that will satisfy the non-negativity conditions, **divide** the right-hand sides of the constraints by the **positive** coefficients of the pivot variable and choose the **smallest** ratio.*

The row that gives the smallest ratio is called the **pivot row**. The element that is in both the pivot row and the pivot column is called the **pivot**.

To complete this iteration the changes in x_2 and x_4 must be reflected into the other constraints and the objective function. This is done by :

- Dividing through constraint (2) by the coefficient of x_2 *i.e.*

$$(2) \Rightarrow 1/2 x_1 + x_2 + 1/2 x_4 = 450 \quad \text{---- } (2')$$

- Transposing this expression to make x_2 the subject *i.e.*

$$(2') \Rightarrow x_2 = 450 - 1/2 x_1 - 1/2 x_4 \quad \text{---- } (5)$$

- Substituting the expression for x_2 into the other constraints and the objective function *i.e.*

$$(1) \& (5) \Rightarrow 3x_1 + 4(450 - 1/2 x_1 - 1/2 x_4) + x_3 = 2400$$

$$i.e. \quad x_1 + x_3 - 2x_4 = 600 \quad \text{---- } (1')$$

$$(3) \& (5) \Rightarrow 2x_1 + 3(450 - 1/2 x_1 - 1/2 x_4) + x_5 = 1600$$

$$i.e. \quad 1/2 x_1 - 3/2 x_4 + x_5 = 250 \quad \text{---- } (3')$$

$$z - 5x_1 - 9(450 - 1/2 x_1 - 1/2 x_4) = 0$$

$$i.e. \quad z - 1/2 x_1 + 9/2 x_4 = 4050$$

Combining these expressions the linear programming problem becomes :

$$\begin{array}{ll}
\text{Minimise} & z - 1/2\,x_1 + 9/2\,x_4 = 4050 \\
\text{Subject to :} & x_1 + x_3 - 2x_4 = 600 \qquad \text{---- } (1') \\
& 1/2\,x_1 + x_2 + 1/2\,x_4 = 450 \qquad \text{---- } (2') \\
& 1/2\,x_1 - 3/2\,x_4 + x_5 = 250 \qquad \text{---- } (3') \\
& 2x_1 + x_6 = 1200 \qquad \text{---- } (4') \\
& x_1, x_2, x_3, x_4, x_5, x_6 \geq 0
\end{array}$$

Notice that during this process the coefficients of the pivot variable x_2 have become a column of the **identity matrix** *i.e.* $\begin{bmatrix} 0 & 0 & 1 & 0 & 0 \end{bmatrix}^{\mathrm{T}}$.

Key Point

To reflect the new values of the variables into the problem, divide through the pivot row by the pivot to make the pivot into a one and then perform elementary row operations to put zeros in all other positions in the pivot column.

This process is called **Jordan elimination**.

A basic feasible solution of this problem is :

$$x_1 = x_4 = 0, x_2 = 450, x_3 = 600, x_5 = 250, x_6 = 1200 \text{ and } z = 4050$$

However, this is **not** the optimal solution of the problem. In the objective function the coefficient of x_1 is negative *i.e.* increasing x_1 will reduce the value of z further. To find the optimal solution this procedure must be applied again.

The options for increasing x_1 are :

a. $(1') \Rightarrow$ increase x_1 to $600 \div 1$ *i.e.* to 600 and reduce x_3 to 0

b. $(2') \Rightarrow$ increase x_1 to $450 \div 1/2$ *i.e.* to 900 and reduce x_2 to 0

c. $(3') \Rightarrow$ increase x_1 to $250 \div 1/2$ *i.e.* to 500 and reduce x_5 to 0

d. $(4') \Rightarrow$ increase x_1 to $1200 \div 2$ *i.e.* to 600 and reduce x_6 to 0

To ensure that the new values of the variables satisfy the non-negativity conditions option **c** must be chosen. Increasing x_1 beyond 500 will make one or more of the basic variables negative (*e.g.* x_5).

Reflecting these changes into the constraints and the objective function using Jordan elimination :

$$(3') \Rightarrow x_1 - 3x_4 + 2x_5 = 500 \quad \text{---- } (3'')$$

$$(3'') \Rightarrow x_1 = 500 + 3x_4 - 2x_5 \quad \text{---- } (6)$$

$$(1') \,\&\, (6) \Rightarrow \left(500 + 3x_4 - 2x_5\right) + x_3 - 2x_4 = 600$$

$$\textit{i.e.} \quad x_3 + x_4 - 2x_5 = 100 \quad \text{----} \,(1'')$$

$$(2') \,\&\, (6) \Rightarrow 1/2\left(500 + 3x_4 - 2x_5\right) + x_2 + 1/2x_4 = 450$$

$$\textit{i.e.} \quad x_2 + 2x_4 - x_5 = 200 \quad \text{----} \,(2'')$$

$$(4') \,\&\, (6) \Rightarrow 2\left(500 + 3x_4 - 2x_5\right) + x_6 = 1200$$

$$\textit{i.e.} \quad 6x_4 - 4x_5 + x_6 = 200 \quad \text{----} \,(4'')$$

$$z - 1/2\left(500 + 3x_4 - 2x_5\right) + 9/2x_4 = 4050$$

$$\textit{i.e.} \quad z + 3x_4 + x_5 = 4300$$

Combining these expressions the linear programming problem becomes :

$$
\begin{aligned}
\text{Minimise} \quad & z + 3x_4 + x_5 = 4300 \\
\text{Subject to :} \quad & x_3 + x_4 - 2x_5 = 100 && \text{----} \,(1'') \\
& x_2 + 2x_4 - x_5 = 200 && \text{----} \,(2'') \\
& x_1 - 3x_4 + 2x_5 = 500 && \text{----} \,(3'') \\
& 6x_4 - 4x_5 + x_6 = 200 && \text{----} \,(4'') \\
& x_1, x_2, x_3, x_4, x_5, x_6 \geq 0
\end{aligned}
$$

In the objective function the coefficients of x_4 and x_5 are positive *i.e.* increasing their values will only **increase** the value of z. Hence, the optimal solution of the problem has been found.

Key Point
*The iterations continue until all of the coefficients in the objective function are **greater than or equal to zero**.*

During the solution x_4 and x_5 were reduced to zero. Hence, from the objective function and constraints $(2'')$ and $(3'')$ the optimal solution of the problem is :

$$x_2 = 200, \ x_1 = 500 \ \text{ and } \ z = 4300$$

Although this procedure may seem to be complicated the key points identified in the example above can be combined to produce a procedure that is relatively easy to use.

Note
The algebraic method is **NOT** used to solve linear programming problems in practice.

Summary of the Simplex Method
The simplex method can be summarised as follows :

Step 1
Write the linear programming problem in **canonical form**. It is usual to write the standardised problem in the form of a **tableau** (*i.e.* table).

Step 2
Examine the coefficients in the objective function. If $c_i \geq 0 \ \forall i$ then **stop**. The optimal solution of the problem has been found.

Step 3
Find the variable that will reduce z by the largest amount. This is the variable that has the largest negative coefficient in the objective function. This variable is called the **pivot variable** and the column containing this variable in the tableau is called the **pivot column**. Suppose that the pivot column is **column k.**

Step 4
Find the largest increase in the pivot variable that will satisfy the non-negativity conditions. To do this, divide the right-hand sides of the constraints by the **positive** coefficients in the pivot column and choose the smallest ratio. The row containing the smallest ratio is called the **pivot row** and the element that is in both the pivot row and the pivot column is called the **pivot**. Suppose that the pivot row is **row p.** Then, in terms of the general linear programming problem given earlier, the pivot row is chosen so that :

$$\frac{b_p}{a_{pk}} = \min\left\{ \frac{b_i}{a_{ik}} \mid a_{ik} > 0 \ ; \ 1 \leq i \leq m \right\}$$

This formula is called the **pivot selection rule**.

Step 5
Use **Jordan elimination** to reflect the changes in the variable values into the constraints and the objective function *i.e.*

- Divide through the pivot row by the pivot to make the pivot into a <u>one</u>.

- Add/subtract multiples of the pivot row to/from the other rows in the tableau to eliminate the pivot variable from the other constraints and the objective function *i.e.* to put <u>zero</u>s in all other positions in the pivot column. This operation transforms the pivot column into a column of the identity matrix.

Note
During this process the pivot variable becomes basic and one of the other basic variables becomes non-basic. Hence, the number of basic variables and the number of non-basic variables in the problem remain fixed throughout the solution.

Step 6
Go back to **Step 2**.

Example 1.3
To illustrate the simplex method reconsider the problem :

$$\text{Maximise} \quad z = 5x_1 + 9x_2$$
$$\text{Subject to :} \quad 3x_1 + 4x_2 \leq 2400$$
$$x_1 + 2x_2 \leq 900$$
$$2x_1 + 3x_2 \leq 1600$$
$$2x_1 \leq 1200$$
$$x_1, x_2 \geq 0$$

Step 1
This problem can be written in canonical form as :

$$\text{Minimise} \quad -\bar{z} - 5x_1 - 9x_2 = 0$$
$$\text{Subject to :} \quad 3x_1 + 4x_2 + x_3 = 2400 \qquad \text{---- (1)}$$
$$x_1 + 2x_2 + x_4 = 900 \qquad \text{---- (2)}$$
$$2x_1 + 3x_2 + x_5 = 1600 \qquad \text{---- (3)}$$
$$2x_1 + x_6 = 1200 \qquad \text{---- (4)}$$
$$x_1, x_2, x_3, x_4, x_5, x_6 \geq 0$$

The initial tableau is :

$-\bar{z}$	x_1	x_2	x_3	x_4	x_5	x_6	**RHS**
1	-5	-9	0	0	0	0	0
0	3	4	1	0	0	0	2400
0	1	2	0	1	0	0	900
0	2	3	0	0	1	0	1600
0	2	0	0	0	0	1	1200

Notice that the columns associated with the **basic variables** in the tableau are columns of the **identity matrix**. This is a useful way of identifying the basic variables as the method proceeds.

The initial basic feasible solution of the problem is :

$$x_1 = x_2 = 0, x_3 = 2400, x_4 = 900, x_5 = 1600, x_6 = 1200 \text{ and } z = 0$$

Step 2
The coefficients of x_1 and x_2 in the objective function are negative *i.e.* the solution above is **not** the optimal solution of the problem.

Step 3

The largest negative coefficient in the objective function is -9 *i.e.*

Hence, the pivot variable is x_2.

Step 4

The smallest ratio in the tableau is 450 *i.e.*

Hence, the pivot is the **2** in the constraint (2) row and the x_2 column.

Step 5

Applying Jordan elimination :

$-\overline{z}$	x_1	x_2	x_3	x_4	x_5	x_6	RHS
1	$-1/2$	0	0	$9/2$	0	0	4050
0	1	0	1	-2	0	0	600
0	$1/2$	1	0	$1/2$	0	0	450
0	$1/2$	0	0	$-3/2$	1	0	250
0	2	0	0	0	0	1	1200

From this tableau a basic feasible solution of the problem is :

$$x_1 = x_4 = 0, x_2 = 450, x_3 = 600, x_5 = 250, x_6 = 1200 \text{ and } z = 4050$$

Step 6
Go back to **Step 2**.

Step 2
The coefficient of x_1 in the objective function is negative *i.e.* the solution above is **not** the optimal solution of the problem.

Step 3/Step 4
The pivot is the **1/2** in the constraint (3) row and the x_1 column *i.e.*

	Pivot Row							Smallest Ratio

Pivot Variable —

Largest Negative Coefficient

$-z$	x_1	x_2	x_3	x_4	x_5	x_6	RHS	
1	$-1/2$	0	0	9/2	0	0	4050	**Ratio**
0	1	0	1	-2	0	0	600	600/1=600
0	1/2	1	0	1/2	0	0	450	450/0.5=900
0	1/2	0	0	$-3/2$	1	0	250	250/0.5=500
0	2	0	0	0	0	1	1200	1200/2=600

Pivot Column —

Pivot

Step 5
Applying Jordan elimination :

$-z$	x_1	x_2	x_3	x_4	x_5	x_6	RHS
1	0	0	0	3	1	0	4300
0	0	0	1	1	-2	0	100
0	0	1	0	2	-1	0	200
0	1	0	0	-3	2	0	500
0	0	0	0	6	-4	1	200

From this tableau a basic feasible solution of the problem is :

$$x_1 = 500, x_2 = 200, x_3 = 100, x_4 = x_5 = 0, x_6 = 200 \text{ and } z = 4300$$

Step 6
Go back to **Step 2**.

Step 2
All the coefficients in the objective function are ≥ 0 *i.e.* the solution above is the optimal solution of the problem.

1.11 Exercises 1b

1. In the context of linear programming explain what is meant by :

    (i)   A basic solution.
    (ii)   A non-basic variable.
    (iii) A basic variable.
    (iv)   A basic feasible solution.
    (v)   A slack variable.
    (vi)   Canonical form.

2. Write the following linear programming problem in canonical form :

    $$\text{Maximise} \quad z = 4x_1 - 3x_2 + 7x_3 + 2x_4 - x_5 - 125$$
    $$\text{Subject to :} \quad x_1 - x_3 + x_4 - 7x_5 \leq 6$$
    $$6x_1 - x_2 + x_4 - 4x_5 \geq -7$$
    $$5x_1 + 2x_3 - 7x_4 \leq 8$$
    $$x_2 - 8x_3 + x_4 - 6x_5 \leq 7$$
    $$x_1, x_2, x_3, x_4, x_5 \geq 0$$

3. (i)   Use the simplex method to solve the following linear programming problem :

    $$\text{Maximise} \quad z = 4x_1 + 3x_2$$
    $$\text{Subject to :} \quad x_1 + x_2 \leq 40$$
    $$2x_1 + x_2 \leq 50$$
    $$x_1, x_2 \geq 0$$

    (ii) Sketch the feasible region for the linear programming problem in 3(i). Annotate your diagram to show the path taken by the simplex method as it converges to the optimal solution.

4. Fortesques's of London is a high class grocer that makes and sells two blends of coffee *i.e.* Breakfast Blend, an ordinary blend and After Dinner, a special blend. Each blend is produced using three types of beans *i.e.* Arabica, Blue Mountain and Costa Rica. The ordinary blend is 1 part Arabica, 3 parts Blue Mountain and 3 parts Costa Rica while the special blend is 2 parts Arabica, 2 parts Blue Mountain and 1 part Costa Rica. The coffee importers are able to supply Fortesques's with up to $120kg$ of Arabica, $180kg$ of Blue Mountain and $150kg$ of Costa Rica each week. The profit made on the ordinary blend is $25p$ per kg and on the special blend is $50p$ per kg. Fortesque's blends are in such demand that the shop can sell all of the coffee it produces.

(i) Formulate a linear programming model from which maximum weekly profit can be calculated.

(ii) Solve your linear programming model from 4(i) using the simplex method. Interpret the solution you obtain.

5. A company manufactures two types of rope called Domestic and Heavy Duty using three different grades of nylon. Table 1 shows the amount of each grade of nylon (in grams) required to make one metre of each type of rope :

Nylon	Rope	
	Domestic	Heavy Duty
Grade 1	3	6
Grade 2	4	7
Grade 3	5	4

Table 1

The company makes 40p per metre profit on Domestic rope and 25p per metre profit on Heavy Duty rope. The weekly availability (in grams) of each grade of nylon is shown in Table 2.

Nylon	Availability
Grade 1	1100
Grade 2	1900
Grade 3	1400

Table 2

The Works Manager wishes to calculate the amount of each type of rope the company should manufacture in order to maximise its weekly profit.

(i) Formulate a linear programming model from which the maximum weekly profit can be calculated.

(ii) Solve your linear programming model from 5(i) using the simplex method. Interpret the solution you obtain.

(iii) Use your solution from 5(ii) to determine how much of each grade of nylon is unused.

6. An oil company can buy two grades of crude oil called Light and Heavy respectively. At most 20 loads of Light and 30 loads of Heavy are available each day. Each load of Light costs £100 and each load of Heavy costs £120. The Heavy can be refined into Diesel at a cost of £25 per load. One load of Heavy produces one load of Diesel and each load of Diesel sells for £200. The Light can be preprocessed into Heavy at a cost of £25 per load. One load of Light produces one load of Heavy. Alternatively, the Light can be refined into Petrol at a cost of £30 per load. One load of Light produces one load of Petrol and each load of Petrol sells for £175. The refinery for Heavy has a maximum capacity of 25 loads per day. Market research suggests that at most 30 loads of fuel *i.e.* Diesel and Petrol combined, can be sold each day.

(i)   Formulate a linear programming model from which the optimal production strategy can be determined *i.e.* how much Light and how much Heavy should be purchased each day and how it should be refined in order to maximise the daily profit.

(ii)   Solve your linear programming model from 6(i) using the simplex method. Interpret the solution you obtain.

2. Solving General Linear Programming Problems

2.1 Introduction

In the description of the simplex method in Chapter 1 the problems considered contained **type 1** (*i.e.* $\leq$) constraints only. However, linear programming problems may also contain **type 2** (*i.e.* =) and **type 3** (*i.e.* $\geq$) constraints. In this chapter the basic algorithm will now be extended to deal with general problems of this kind.

2.2 Development of the Extended Algorithm

Consider the following linear programming problem :

$$\begin{aligned}
\text{Maximise} \quad & z = 10x_1 + 14x_2 + 11x_3 + x_4 \\
\text{Subject to :} \quad & 0.3x_1 + 0.5x_2 + 0.4x_3 \leq 2000 \quad \text{---- (1)} \\
& 3x_1 + 4x_2 + 3x_3 + x_4 = 15000 \quad \text{---- (2)} \\
& x_1 \geq 1000 \quad \text{---- (3)} \\
& x_1, x_2, x_3, x_4 \geq 0
\end{aligned}$$

The first stage in the solution is to write this problem in canonical form *i.e.* as a minimisation problem in which each constraint is an equation that contains a basic variable with a coefficient of one.

The objective function can be rewritten as :

$$\text{Minimise} \quad -\bar{z} - 10x_1 - 14x_2 - 11x_3 - x_4 = 0$$

Each **type 1** constraint can be converted into an equation by adding a non-negative **slack variable**. Adding a slack variable x_5, constraint (1) can be written as :

$$0.3x_1 + 0.5x_2 + 0.4x_3 + x_5 = 2000$$

Each **type 2** constraint is already an equation.

Each **type 3** constraint can be converted into an equation by subtracting a non-negative **surplus variable**. If the constraint represents the supply of a commodity then the surplus variable will represent the excess supply of that commodity. Subtracting a surplus variable x_6, constraint (3) can be written as :

$$x_1 - x_6 = 1000$$

Combining these expressions the linear programming problem becomes :

$$\begin{aligned}
\text{Minimise} \quad & -\bar{z} - 10x_1 - 14x_2 - 11x_3 - x_4 = 0 \\
\text{Subject to :} \quad & 0.3x_1 + 0.5x_2 + 0.4x_3 + x_5 = 2000 \quad \text{---- (1')} \\
& 3x_1 + 4x_2 + 3x_3 + x_4 = 15000 \quad \text{---- (2')} \\
& x_1 - x_6 = 1000 \quad \text{---- (3')} \\
& x_1, x_2, x_3, x_4, x_5, x_6 \geq 0
\end{aligned}$$

29

Although this is a minimisation problem and all of the constraints are equations, this problem is **not** in canonical form. Constraints $(2')$ and $(3')$ do not contain a basic variable with a coefficient of one. To overcome this problem a new type of non-negative variable called an **artificial variable** is added to each **type 2** and **type 3** constraint.

Note

It is important to be able to identify the artificial variables during the solution and so they are usually given distinguishing names such as $a_1, a_2, a_3, \ldots$ or $x_7^*, x_8^*, x_9^*, \ldots$ etc.

Adding artificial variables a_1 and a_2 to constraints $(2')$ and $(3')$ respectively the linear programming problem can be written as :

$$\begin{aligned} \text{Minimise} \quad & -\bar{z} - 10x_1 - 14x_2 - 11x_3 - x_4 = 0 \\ \text{Subject to :} \quad & 0.3x_1 + 0.5x_2 + 0.4x_3 + x_5 = 2000 \quad \text{---- } (1'') \\ & 3x_1 + 4x_2 + 3x_3 + x_4 + a_1 = 15000 \quad \text{---- } (2'') \\ & x_1 - x_6 + a_2 = 1000 \quad \text{---- } (3'') \\ & x_1, x_2, x_3, x_4, x_5, x_6, a_1, a_2 \geq 0 \end{aligned}$$

This problem **is** now in canonical form. However, adding the artificial variables a_1 and a_2 has created a **new** linear programming problem (since the artificial variables were added to one side of the constraints but not to the other). The solution of the new problem may **not** be the solution of the original problem. In particular, the initial solution of the new problem may not be the initial solution of the original problem. The simplex method requires an initial basic feasible solution of the original problem before it can begin to find the optimal solution.

The new problem **will** have the same solution as the original problem if the optimal values of the artificial variables are **zero**. Hence, this problem can be overcome by using the simplex method to drive the values of the artificial variables to zero. This process also produces the initial basic feasible solution required.

The procedure used is as follows :

- Define a new objective function w to be the sum of the artificial variables.

- Add this objective function to the new linear programming problem. The resulting linear programming problem will not be in canonical form (since the constraints that contain artificial variables will not contain a basic variable with a coefficient of one). To overcome this problem, add the constraints that contain artificial variables and then subtract this expression from the w-row.

- Apply the simplex method to the w-row. This gives the minimum value of w i.e. w_{min}. Then :

    - If $w_{min} = 0$, all of the artificial variables must be zero (since they are all constrained to be non-negative). In this case the simplex method will have found the initial basic feasible solution required. The simplex method can then be applied to the z-row to find the optimal solution of the problem.

    - If $w_{min} > 0$, at least one of the artificial variables must be greater than zero. In this case the simplex method will have failed to find an equivalent linear programming problem to the original one. In this case the original linear programming problem is said to be **infeasible**.

This extended procedure is called the **two-phase simplex method**. Here :

Phase 1 involves minimising the value of w to obtain an initial basic feasible solution *i.e.* phase 1 involves establishing **feasibility**. During this phase the z-row plays a purely passive role *i.e.* it is transformed by Jordan elimination but is **never** selected as the pivot row.

Phase 2 involves minimising the value of z to obtain the optimal solution *i.e.* phase 2 involves achieving **optimality**. During this phase the w-row is treated as a constraint and $(-w)$ is treated as a non-negative variable *i.e.* the w-row is transformed by Jordan elimination and **can** be selected as the pivot row.

Notes
- Once the artificial variables become non-basic they can be discarded from the phase 1 tableau. This prevents them from being selected as pivot variables at a later stage *i.e.* from becoming basic again. If the artificial variables are basic but zero at the end of phase 1 they must be included in the phase 2 tableau but they can be discarded as soon as they become non-basic.

- If **all** of the artificial variables have been discarded from the phase 1 tableau, the w-row will be **trivial** and can also be discarded.

For the example problem :

$$w = a_1 + a_2 \quad i.e. \quad -w + a_1 + a_2 = 0$$

Adding this expression to the linear programming problem :

$$\begin{aligned}
\text{Minimise} \quad & -\bar{z} - 10x_1 - 14x_2 - 11x_3 - x_4 = 0 \\
& -w + a_1 + a_2 = 0 \quad ---- (1''') \\
\text{Subject to :} \quad & 0.3x_1 + 0.5x_2 + 0.4x_3 + x_5 = 2000 \quad ---- (2''') \\
& 3x_1 + 4x_2 + 3x_3 + x_4 + a_1 = 15000 \quad ---- (3''') \\
& x_1 - x_6 + a_2 = 1000 \quad ---- (4''') \\
& x_1, x_2, x_3, x_4, x_5, x_6, a_1, a_2 \geq 0
\end{aligned}$$

Notice that constraints $(3''')$ and $(4''')$ no longer contain a basic variable with a coefficient of one *i.e.* that the problem is no longer in canonical form. Adding these constraints :

$$(3''') + (4''') \Rightarrow 4x_1 + 4x_2 + 3x_3 + x_4 - x_6 + a_1 + a_2 = 16000 \quad ---- (5''')$$

Subtracting this expression from the w-row :

$$(1''') - (5''') \Rightarrow -w - 4x_1 - 4x_2 - 3x_3 - x_4 + x_6 = -16000$$

Replacing the old w-row with the new one the linear programming problem can be written as :

$$\begin{aligned}
\text{Minimise} \quad & -\bar{z} - 10x_1 - 14x_2 - 11x_3 - x_4 = 0 \\
& -w - 4x_1 - 4x_2 - 3x_3 - x_4 + x_6 = -16000 \\
\text{Subject to :} \quad & 0.3x_1 + 0.5x_2 + 0.4x_3 + x_5 = 2000 \\
& 3x_1 + 4x_2 + 3x_3 + x_4 + a_1 = 15000 \\
& x_1 - x_6 + a_2 = 1000 \\
& x_1, x_2, x_3, x_4, x_5, x_6, a_1, a_2 \geq 0
\end{aligned}$$

This problem **is** now in canonical form.

The **phase 1** simplex tableau is :

$-\overline{z}$	$-w$	x_1	x_2	x_3	x_4	x_5	x_6	a_1	a_2	RHS
1	0	-10	-14	-11	-1	0	0	0	0	0
0	1	-4	-4	-3	-1	0	1	0	0	-16000
0	0	3/10	1/2	2/5	0	1	0	0	0	2000
0	0	3	4	3	1	0	0	1	0	15000
0	0	1	0	0	0	0	-1	0	1	1000
1	0	0	-14	-11	-1	0	-10	0	10	10000
0	1	0	-4	-3	-1	0	-3	0	4	-12000
0	0	0	1/2	2/5	0	1	3/10	0	-3/10	1700
0	0	0	4	3	1	0	3	1	-3	12000
0	0	1	0	0	0	0	-1	0	1	1000
1	0	0	0	-1/2	5/2	0	1/2	7/2		52000
0	1	0	0	0	0	0	0	1		0
0	0	0	0	1/40	-1/8	1	-3/40	-1/8		200
0	0	0	1	3/4	1/4	0	3/4	1/4		3000
0	0	1	0	0	0	0	-1	0		1000

Since all of the coefficients in the w-row are now greater than or equal to zero phase 1 is complete. From the final tableau it can be seen that $w_{min} = 0$ *i.e.* the original problem is **feasible** and phase 2 can begin.

Notes
- Notice that the artificial variables a_1 and a_2 have become non-basic. These variables can be discarded from the phase 2 tableau. The w-row is now trivial and can also be discarded.

- From the final tableau above it can be seen that the initial basic feasible solution of the original problem is $x_1 = 1000$, $x_2 = 3000$, $x_3 = 0$, $x_4 = 0$ and $z = 52000$.

The **phase 2** simplex tableau is :

$-\overline{z}$	x_1	x_2	x_3	x_4	x_5	x_6	RHS
1	0	0	-1/2	5/2	0	1/2	52000
0	0	0	1/40	-1/8	1	-3/40	200
0	0	1	3/4	1/4	0	3/4	3000
0	1	0	0	0	0	-1	1000
1	0	2/3	0	8/3	0	1	54000
0	0	-1/30	0	-2/15	1	-1/10	100
0	0	4/3	1	1/3	0	1	4000
0	1	0	0	0	0	-1	1000

Since all of the coefficients in the z-row are now greater than or equal to zero phase 2 is complete and the solution is **optimal**. From the final tableau the optimal solution of the linear programming problem is :

$$x_1 = 1000, \ x_2 = 0, \ x_3 = 4000, \ x_4 = 0 \text{ and } z = 54000$$

2.3 Summary of the Two-Phase Simplex Method
The two-phase simplex method can be summarised as follows :

Step 1
Write the linear programming problem in **canonical form** *i.e.*

- Convert the problem into a minimisation problem (if necessary).

- Add a **slack variable** to each **type 1** constraint.

- Add an **artificial variable** to each **type 2** constraint.

- Subtract a **surplus variable** <u>and</u> add an **artificial variable** to each **type 3** constraint.

- Form a new objective function w that is defined to be the sum of the artificial variables.

- Add the constraints that contain artificial variables and then subtract this expression from the w-row.

- Write down the new linear programming problem *i.e.* the z-row, the w-row, the constraints and the non-negativity conditions (that now include the slack, surplus and artificial variables).

Step 2
Complete **phase 1** *i.e.* apply the simplex method to the w-row. This gives the minimum value of w *i.e.* w_{min}. Then :

- If $w_{min} = 0$, the simplex method will have found an initial basic **feasible** solution of the problem and phase 2 can be completed. Go to **Step 3**.

- If $w_{min} > 0$, the original linear programming problem is **infeasible** and the procedure <u>stops</u>.

During this phase the z-row plays a purely passive role *i.e.* it is transformed by Jordan elimination but is <u>**never**</u> selected as the pivot row.

Step 3
Complete **phase 2** *i.e.* apply the simplex method to the z-row to find the **optimal** solution of the problem.

During this phase the w-row is treated as a constraint and $(-w)$ is treated as a non-negative variable *i.e.* the w-row is transformed by Jordan elimination and <u>**can**</u> be selected as the pivot row.

Notes
- Once the artificial variables become non-basic they can be discarded from the phase 1 tableau. If the artificial variables are basic but zero at the end of phase 1 they must be included in the phase 2 tableau but they can be discarded as soon as they become non-basic.

- If <u>**all**</u> of the artificial variables have been discarded from the phase 1 tableau, the w-row will be trivial and can also be discarded.

2.4 A Complete Example

Consider the following linear programming problem :

$$\begin{aligned}
\text{Minimise} \quad & z = 2x_1 - x_2 \\
\text{Subject to :} \quad & x_2 \leq 1 \quad \text{---- (1)} \\
& x_1 \leq 1 \quad \text{---- (2)} \\
& x_1 - x_2 = 1/2 \quad \text{---- (3)} \\
& x_1 + x_2 \geq 1 \quad \text{---- (4)} \\
& x_1, x_2 \geq 0
\end{aligned}$$

The objective function can be written as :

$$-z + 2x_1 - x_2 = 0$$

Adding slack variables x_3 and x_4 to constraints (1) and (2) respectively :

$$(1) \Rightarrow x_2 + x_3 = 1 \quad \text{---- (1')}$$
$$(2) \Rightarrow x_1 + x_4 = 1 \quad \text{---- (2')}$$

Adding an artificial variable a_1 to constraint (3) :

$$(3) \Rightarrow x_1 - x_2 + a_1 = 1/2 \quad \text{---- (3')}$$

Subtracting a surplus variable x_5 and adding an artificial variable a_2 to constraint (4) :

$$(4) \Rightarrow x_1 + x_2 - x_5 + a_2 = 1 \quad \text{---- (4')}$$

Let $w = a_1 + a_2$ i.e.

$$-w + a_1 + a_2 = 0 \quad \text{---- (5)}$$

Adding the constraints that contain artificial variables :

$$(3') + (4') \Rightarrow 2x_1 - x_5 + a_1 + a_2 = 3/2 \quad \text{---- (6)}$$

Subtracting this expression from the w-row :

$$(5) - (6) \Rightarrow -w - 2x_1 + x_5 = -3/2$$

Combining these expressions, the linear programming problem in canonical form becomes :

Minimise $\quad -z + 2x_1 - x_2 = 0$

$\qquad\qquad -w - 2x_1 + x_5 = -3/2$

Subject to : $\quad x_2 + x_3 = 1 \quad \text{----} (1')$

$\qquad\qquad x_1 + x_4 = 1 \quad \text{----} (2')$

$\qquad\qquad x_1 - x_2 + a_1 = 1/2 \quad \text{----} (3')$

$\qquad\qquad x_1 + x_2 - x_5 + a_2 = 1 \quad \text{----} (4')$

$\qquad\qquad x_1, x_2, x_3, x_4, x_5, a_1, a_2 \geq 0$

The **phase 1** simplex tableau is :

$-z$	$-w$	x_1	x_2	x_3	x_4	x_5	a_1	a_2	**RHS**
1	0	2	-1	0	0	0	0	0	0
0	1	-2	0	0	0	1	0	0	$-3/2$
0	0	0	1	1	0	0	0	0	1
0	0	1	0	0	1	0	0	0	1
0	0	1	-1	0	0	0	1	0	1/2
0	0	1	1	0	0	-1	0	1	1
1	0	0	1	0	0	0	-2	0	-1
0	1	0	-2	0	0	1	2	0	$-1/2$
0	0	0	1	1	0	0	0	0	1
0	0	0	1	0	1	0	-1	0	1/2
0	0	1	-1	0	0	0	1	0	1/2
0	0	0	2	0	0	-1	-1	1	1/2
1	0	0	0	0	0	1/2		$-1/2$	$-5/4$
0	1	0	0	0	0	0		1	0
0	0	0	0	1	0	1/2		$-1/2$	3/4
0	0	0	0	0	1	1/2		$-1/2$	1/4
0	0	1	0	0	0	$-1/2$		1/2	3/4
0	0	0	1	0	0	$-1/2$		1/2	1/4

Since all of the coefficients in the w-row are now greater than or equal to zero, phase 1 is complete. From the final tableau it can be seen that $w_{\min} = 0$ *i.e.* the original problem is **feasible** and phase 2 can begin.

The artificial variables a_1 and a_2 have become non-basic. These variables can be discarded from the phase 2 tableau. The w-row is trivial and can also be discarded.

The **phase 2** simplex tableau is :

$-z$	x_1	x_2	x_3	x_4	x_5	**RHS**
1	0	0	0	0	1/2	$-5/4$
0	0	0	1	0	1/2	3/4
0	0	0	0	1	1/2	1/4
0	1	0	0	0	$-1/2$	3/4
0	0	1	0	0	$-1/2$	1/4

Since all of the coefficients in the z-row are greater than or equal to zero this tableau is **optimal** *i.e.* no calculations are required in phase 2 in this case. From the final tableau the optimal solution of the linear programming problem is :

$$x_1 = 3/4,\ x_2 = 1/4\ \text{and}\ z = 5/4$$

2.5 Exercises 2

1. Use the two-phase simplex method to solve the following linear programming problems :

   (i)   Maximise $z = 5x_1 + 2x_2 + 2x_3 - 7$

         Subject to : $x_1 + x_2 + x_3 = 1$

                      $x_1 + 2x_2 - x_3 = 2$

                      $x_1, x_2, x_3 \geq 0$

   (ii) Maximise $z = x_1 + x_2$

         Subject to : $4x_1 + 5x_2 \leq 200$

                      $x_1 + 4x_2 \geq 40$

                      $2x_1 + 3x_2 = 90$

                      $x_1, x_2 \geq 0$

   (iii) Minimise $z = 3x_2 - x_3 + 8$

         Subject to : $x_1 + x_2 + x_3 = 10$

                      $2x_1 + 3x_2 + x_3 = 15$

                      $x_1, x_2, x_3 \geq 0$

   (iv) Minimise $z = -x_1 - 2x_2 - x_3$

         Subject to : $-x_1 + 2x_2 + x_3 = 4$

                      $-2x_1 - x_3 \geq 3$

                      $3x_1 + 2x_2 - x_3 \leq 8$

                      $x_1, x_2, x_3 \geq 0$

2. A furniture company manufactures three types of kitchen chair, the Amber, the Beaton and the Countess. The seat of each chair is covered in fabric. This is no longer available from the supplier but there is 2000m left in stock. The company has 15000 lengths of wood available for making kitchen chairs. Any wood left over is sold for £1 per length. The company has an order for 1000 Amber chairs that must be satisfied. Table 1 shows the material requirements for each chair together with the profit made on each chair :

	Model		
	Amber	Beaton	Countess
Amount of fabric per chair (m)	0.3	0.5	0.4
Lengths of wood per chair	3	4	3
Profit per chair (£)	10	14	11

Table 1

(i) Formulate a linear programming model from which the manager of the company can determine the number of each chair to manufacture in order to maximise the total profit.

(ii) Solve your linear programming model from 2(i) using the two-phase simplex method.

(iii) Use your solution from 2(ii) to determine the amount of additional profit the company can make by selling off the unused wood.

(iv) Suppose that after solving your linear programming model from 2(i) it was discovered that a leaking roof caused 100m of the fabric to become damaged. Would your solution from 2(ii) still be feasible ?

3. The Majestic Hotel employs Security Officers for twenty-four hours each day. The working day at the hotel is divided into six four-hour shifts starting at midnight. Each Security Officer works two <u>consecutive</u> shifts (that can be shift six followed by shift one). Table 2 shows the shift times and the minimum number of Security Officers that must be present during each shift in order to meet safety requirements :

Shift	Times	Minimum Number of Security Officers
1	00,00-04.00	4
2	04.00-08.00	4
3	08.00-12.00	6
4	12.00-16.00	8
5	16.00-20.00	6
6	20.00-00.00	4

Table 2

(i) Formulate a linear programming model from which the manager of the hotel can determine the minimum number of Security Officers to employ. Hint : Let x_1 be the number of Security Officers working shift one and shift two, x_2 be the number of Security Officers working shift two and shift three, *etc.*

(ii)   Solve your linear programming model from 3(i) using the Solver tool in *Excel*. An example of using this tool is given in Chapter 10.

(iii)   The manager of the hotel decides to introduce a non-social hours rate of pay for any Security Officer who works during shift one. The new rate is 150% of the normal rate and is paid for the whole eight-hour period worked. Formulate a linear programming model from which the manager of the hotel can determine the number of Security Officers to employ during each shift in order to minimise the total wage bill. Do not attempt to solve your model. Hint : Let the normal rate of pay be £C per hour. Then, the non-social hours rate is £1.5C per hour.

4. Explain why it is not possible for a basic variable to become non-basic at one iteration and then to become basic again at the next iteration.

5. Suppose that one of the variables in a linear programming problem is constrained to be non-positive. How could this problem be converted into a standard linear programming problem *i.e.* one in which all of the variables are constrained to be non-negative ?

3. Problem Cases Revisited

3.1 Introduction

In Chapter 1 it was shown how problem cases such as non-unique solutions, infeasible problems and unbounded problems can be recognised when solving linear programming problems using the graphical method. In this chapter it will be shown how these and other problem cases can be recognised when using the simplex methods.

3.2 Non-Unique Solutions

Consider the following linear programming problem :

$$\text{Minimise} \qquad z = -2x_1 - 2x_2$$
$$\text{Subject to :} \qquad -x_1 + x_2 \leq 2$$
$$x_1 + x_2 \leq 6$$
$$1/2\,x_1 - x_2 \leq 3/2$$
$$x_1, x_2 \geq 0$$

Solving this problem using the graphical method :

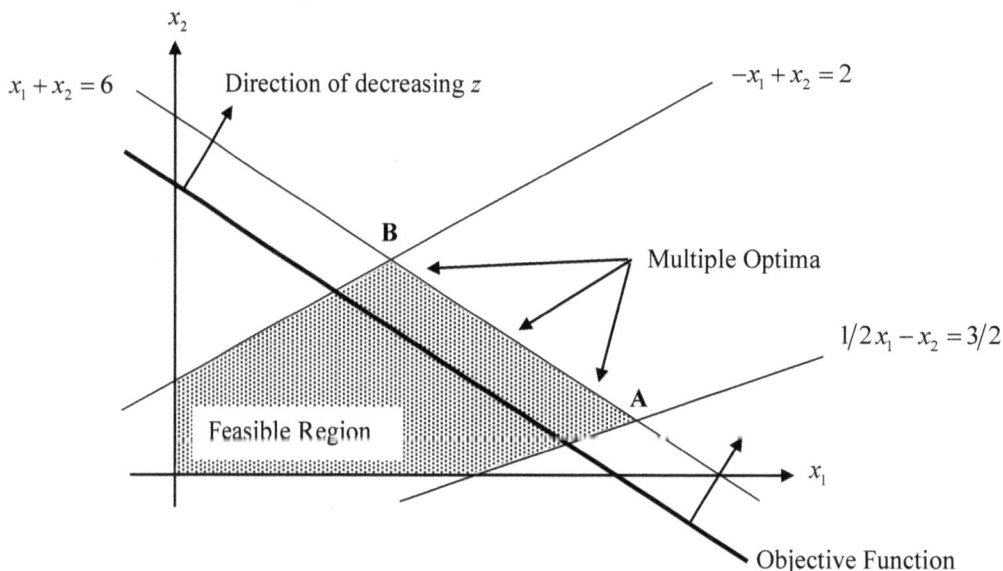

It can be seen that this problem has a non-unique solution. Solving it using the simplex method :

The problem can be written in canonical form as :

$$\text{Minimise} \qquad -z - 2x_1 - 2x_2 = 0$$
$$\text{Subject to :} \qquad -x_1 + x_2 + x_3 = 2$$
$$x_1 + x_2 + x_4 = 6$$
$$1/2\,x_1 - x_2 + x_5 = 3/2$$
$$x_1, x_2, x_3, x_4, x_5 \geq 0$$

The simplex tableau is :

$-z$	x_1	x_2	x_3	x_4	x_5	RHS
1	-2	-2	0	0	0	0
0	-1	1	1	0	0	2
0	1	1	0	1	0	6
0	1/2	-1	0	0	1	3/2
1	0	-6	0	0	4	6
0	0	-1	1	0	2	5
0	0	3	0	1	-2	3
0	1	-2	0	0	2	3
1	0	0	0	2	0	12
0	0	0	1	1/3	4/3	6
0	0	1	0	1/3	-2/3	1
0	1	0	0	2/3	2/3	5

Pivot   Zero Coefficient

By inspection it can be seen that the solution is now optimal. The optimal basic feasible solution is :

$$z = -12, \ x_1 = 5 \text{ and } x_2 = 1$$

This solution corresponds to point **A** on the graph. However, in the optimal tableau the non-basic variable x_5 has a zero coefficient in the z-row. If x_5 is chosen to be the pivot variable and the pivot is chosen to be the greyed value $4/3$ in that column, the following tableau is obtained :

$-z$	x_1	x_2	x_3	x_4	x_5	RHS
1	0	0	0	2	0	12
0	0	0	3/4	1/4	1	9/2
0	0	1	1/2	1/2	0	4
0	1	0	-1/2	1/2	0	2

This tableau is also optimal. The optimal basic feasible solution in this case is :

$$z = -12, \ x_1 = 2 \text{ and } x_2 = 4$$

This solution corresponds to point **B** on the graph. Hence, for this problem there are **two** basic feasible solutions.

Note
In the second optimal tableau the non-basic variable x_3 also has a zero coefficient in the z-row. However, pivoting on this variable simply reproduces the first optimal tableau above.

This example shows how a **non-unique solution** can be recognised within the simplex methods *i.e.*

> A linear programming problem has a **non-unique solution** if in the optimal tableau, a non-basic variable has a zero coefficient in the z-row.

By looking at the graph it can be seen that all points along the boundary of the feasible region between point **A** and point **B** are also solutions of the linear programming problem (although they are **not** necessarily basic feasible solutions). The **general solution** of a linear programming problem of this kind (*i.e.* the equation of the optimal boundary of the feasible region) can be written as a **convex combination** of the optimal basic feasible solutions $\underline{x}_i$ *i.e.* the general solution is an expression in the form :

$$\sum_{i=1}^{t} \alpha_i \underline{x}_i$$

where $\sum_{i=1}^{t} \alpha_i = 1$ and $\alpha_i \geq 0, \quad 1 \leq i \leq t$.

For a two-variable linear programming problem with optimal basic feasible solutions $\underline{x}_1$ and $\underline{x}_2$ this expression can be written as :

$$\alpha \underline{x}_1 + (1-\alpha)\underline{x}_2$$

For example, for the two variable problem above, the optimal basic feasible solutions are :

$$\underline{x}_1 = \begin{bmatrix} 5 \\ 1 \end{bmatrix} \text{ and } \underline{x}_2 = \begin{bmatrix} 2 \\ 4 \end{bmatrix}$$

Hence, the general solution of the linear programming problem is :

$$\alpha \underline{x}_1 + (1-\alpha)\underline{x}_2$$

$$\alpha \begin{bmatrix} 5 \\ 1 \end{bmatrix} + (1-\alpha)\begin{bmatrix} 2 \\ 4 \end{bmatrix}$$

$$\text{i.e.} \quad \begin{bmatrix} 3\alpha+2 \\ 4-3\alpha \end{bmatrix} \quad \forall \alpha \in [0,1]$$

Notice that substituting $\alpha = 1$ into this expression gives the first optimal basic feasible solution and substituting $\alpha = 0$ gives the second optimal basic feasible solution.

3.3 Infeasible Problems

Consider the following linear programming problem :

$$\text{Maximise} \quad z = 3x_1 + 2x_2$$
$$\text{Subject to :} \quad x_1 \leq 2$$
$$-x_1 + x_2 \leq 1$$
$$x_1 + x_2 \geq 8$$
$$x_1, x_2 \geq 0$$

Solving this problem using the graphical method :

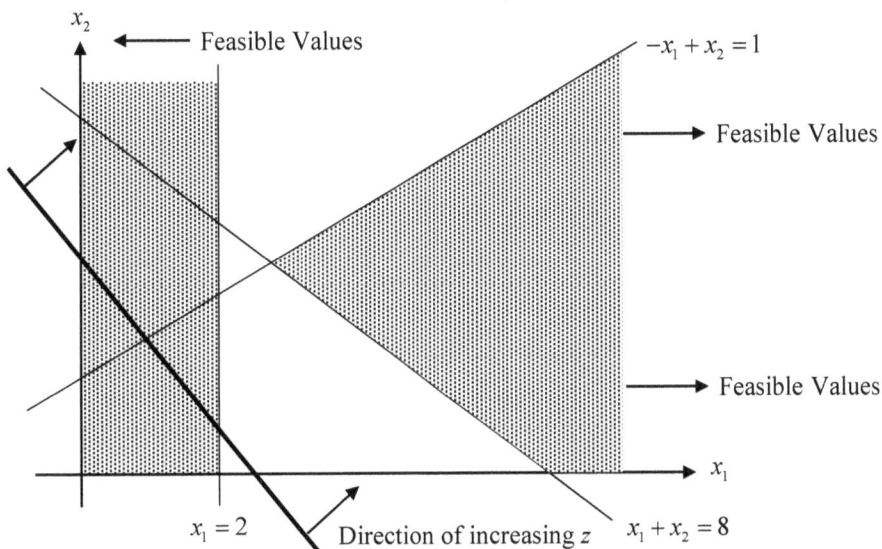

It can be seen that there is no region in the $x_1 - x_2$ plane in which all constraints are satisfied simultaneously. Hence, the feasible region is empty *i.e.* this problem is infeasible. Solving this problem using the simplex method :

The problem can be written in canonical form as :

$$\text{Minimise} \quad -\bar{z} - 3x_1 - 2x_2 = 0$$
$$-w - x_1 - x_2 + x_5 = -8$$
$$\text{Subject to :} \quad x_1 + x_3 = 2$$
$$-x_1 + x_2 + x_4 = 1$$
$$x_1 + x_2 - x_5 + a_1 = 8$$
$$x_1, x_2, x_3, x_4, x_5, a_1 \geq 0$$

The phase 1 simplex tableau is :

$-\bar{z}$	$-w$	x_1	x_2	x_3	x_4	x_5	a_1	RHS
1	0	-3	-2	0	0	0	0	0
0	1	-1	-1	0	0	1	0	-8
0	0	1	0	1	0	0	0	2
0	0	-1	1	0	1	0	0	1
0	0	1	1	0	0	-1	1	8
1	0	-5	0	0	2	0	0	2
0	1	-2	0	0	1	1	0	-7
0	0	1	0	1	0	0	0	2
0	0	-1	1	0	1	0	0	1
0	0	2	0	0	-1	-1	1	7
1	0	0	0	5	2	0	0	12
0	1	0	0	2	1	1	0	-3
0	0	1	0	1	0	0	0	2
0	0	0	1	1	1	0	0	3
0	0	0	0	-2	-1	-1	1	3

It can be seen that phase 1 is now complete with $-w = -3$ *i.e.* $w = 3$ *i.e.* $\text{w}_{min} > 0$

This example shows how an **infeasible problem** can be recognised within the simplex methods *i.e.*

A linear programming problem is **infeasible** if at the end of phase 1, $w_{min} > 0$.

Note
The current **solution** of a linear programming problem is **infeasible** if a negative value appears on the right-hand side of one of the constraints. Feasibility can sometimes be restored using an algorithm called the dual simplex method. This procedure will be described in Chapter 7. If feasibility cannot be restored then the linear programming **problem** itself is **infeasible**.

3.4 Unbounded Problems
Consider the following linear programming problem :

$$\text{Minimise} \quad z = -x_1 - 1/2\,x_2$$
$$\text{Subject to :} \quad -x_1 + x_2 \le 2$$
$$-3x_1 + x_2 \le 1$$
$$x_1 - 2x_2 \le 1$$
$$x_1, x_2 \ge 0$$

Solving this problem using the graphical method :

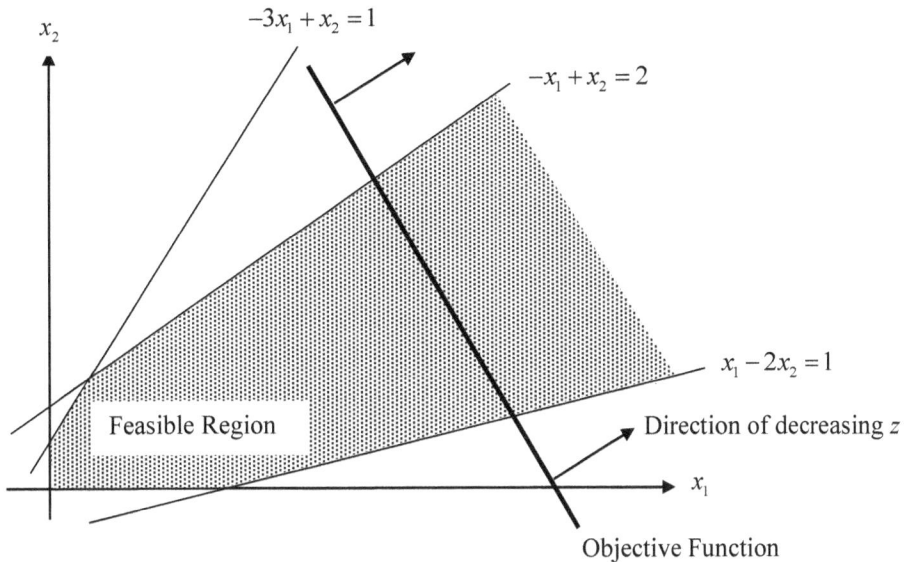

It can be seen that this problem is unbounded. Solving this problem using the simplex method :

The problem can be written in canonical form as :

$$\text{Minimise} \quad -z - x_1 - 1/2 x_2 = 0$$
$$\text{Subject to :} \quad -x_1 + x_2 + x_3 = 2$$
$$-3x_1 + x_2 + x_4 = 1$$
$$x_1 - 2x_2 + x_5 = 1$$
$$x_1, x_2, x_3, x_4, x_5 \geq 0$$

The simplex tableau is :

$-z$	x_1	x_2	x_3	x_4	x_5	**RHS**
1	-1	$-1/2$	0	0	0	0
0	-1	1	1	0	0	2
0	-3	1	0	1	0	1
0	1	-2	0	0	1	1
1	0	$-5/2$	0	0	1	1
0	0	-1	1	0	1	3
0	0	-5	0	1	3	4
0	1	-2	0	0	1	1

Negative Values

It can be seen that the pivot selection rule has broken down. Although x_2 can be selected as the pivot variable for the second iteration the pivot row cannot be found since all of the values in the x_2 column are negative.

A. M. FITZHARRIS

This example shows how an **unbounded problem** can be recognised within the simplex methods *i.e.*

A linear programming problem is **unbounded** if at any stage during the solution the pivot selection rule breaks down *i.e.* if :

$$S = \left\{ \frac{b_i}{a_{ik}} \mid a_{ik} > 0 \; ; \; 1 \le i \le m \right\} = \phi \quad i.e. \text{ the empty set}$$

3.5 Degeneracy
A linear programming problem is said to be **degenerate** if at any stage during the solution a zero appears on the right-hand side of a constraint row. If that constraint row is then chosen to be the pivot row, the next tableau will give the same basic feasible solution. To illustrate this problem consider the following example :

Minimise $\quad z = -x_1 - 2x_2$
Subject to $\quad -2x_1 + 2x_2 \le 3$
$\qquad\qquad 2x_1 + 2x_2 \le 3$
$\qquad\qquad x_1, x_2 \ge 0$

This problem can be written in canonical form as :

Minimise $\quad -z - x_1 - 2x_2 = 0$
Subject to : $\quad -2x_1 + 2x_2 + x_3 = 3$
$\qquad\qquad 2x_1 + 2x_2 + x_4 = 3$
$\qquad\qquad x_1, x_2, x_3, x_4 \ge 0$

The simplex tableau is :

$-z$	x_1	x_2	x_3	x_4	RHS
1	-1	-2	0	0	0
0	-2	2	1	0	3
0	2	2	0	1	3
1	-3	0	1	0	3
0	-1	1	1/2	0	3/2
0	4	0	-1	1	0
1	0	0	1/4	3/4	3
0	0	1	1/4	1/4	3/2
0	1	0	-1/4	1/4	0

Zero Coefficients

It can be seen that in the second tableau a zero has appeared on the right-hand side of a constraint row. This constraint row has then been selected as the pivot row for the next iteration. The second and third tableaux are clearly different. However, from the second tableau the basic feasible solution is :

$$z = -3,\ x_1 = 0 \text{ and } x_2 = 3/2$$

and from the third tableau the optimal basic feasible solution is also :

$$z = -3,\ x_1 = 0 \text{ and } x_2 = 3/2$$

i.e. the tableaux are changing but the basic feasible solution is not.

Degeneracy occurs when there are redundant constraints in the problem. To illustrate this consider the graphical solution of the linear programming problem above :

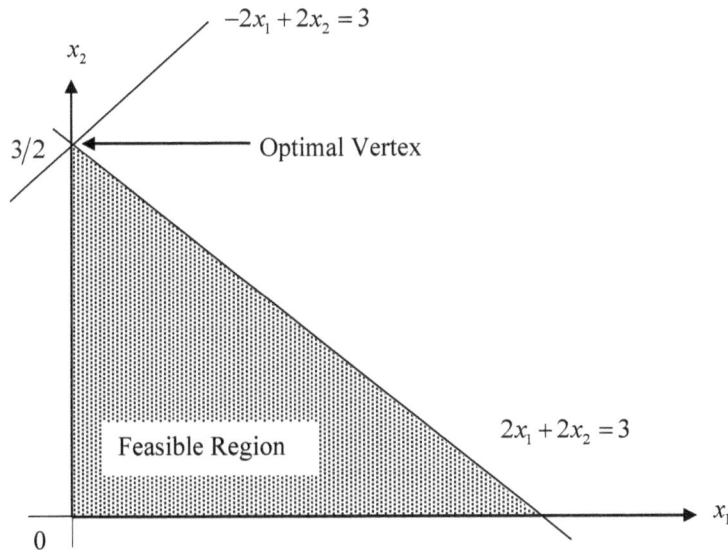

It can be seen that the optimal vertex occurs at the intersection of three lines *i.e.* the line $-2x_1 + 2x_2 = 3$, the line $2x_1 + 2x_2 = 3$ and the x_2 axis *i.e.* the line $x_1 = 0$.

By looking at the canonical form of the problem *i.e.*

$$\begin{aligned} \text{Minimise} \quad & -z - x_1 - 2x_2 = 0 \\ \text{Subject to :} \quad & -2x_1 + 2x_2 + x_3 = 3 \\ & 2x_1 + 2x_2 + x_4 = 3 \\ & x_1, x_2, x_3, x_4 \geq 0 \end{aligned}$$

it can be seen that :

- The lines $x_1 = 0$ and $-2x_1 + 2x_2 = 3$ intersect when $x_1 = 0$ and $x_3 = 0$ *i.e.* when x_1 and x_3 are non-basic.

- The lines $x_1 = 0$ and $2x_1 + 2x_2 = 3$ intersect when $x_1 = 0$ and $x_4 = 0$ *i.e.* when x_1 and x_4 are non-basic.

- The lines $-2x_1 + 2x_2 = 3$ and $2x_1 + 2x_2 = 3$ intersect when $x_3 = 0$ and $x_4 = 0$ *i.e.* when x_3 and x_4 are non-basic.

i.e. that each intersection corresponds to a different set of non-basic (and hence basic) variables *i.e.* that each intersection corresponds to a different tableau *i.e.* that three (different) tableaux produce the same basic feasible solution.

If a linear programming problem is highly degenerate it is possible for the sequence of tableaux to repeat itself. In this case the algorithm will be stuck in a loop and the optimal solution will never be found. This problem is called **cycling**. Cycling can usually be prevented by using a procedure known as **Bland's Rule**.

Bland's Rule

- When selecting the **pivot column**, read from left to right and choose the pivot variable as the <u>first</u> x_k for which $c_k < 0$.

- If there is a choice for the **pivot row**, choose the row whose basic variable has the <u>smallest</u> <u>subscript</u>.

To illustrate the use of Bland's Rule the linear programming problem from the last example *i.e.*

$$\begin{aligned} \text{Minimise} \quad & z = -x_1 - 2x_2 \\ \text{Subject to} \quad & -2x_1 + 2x_2 \leq 3 \\ & 2x_1 + 2x_2 \leq 3 \\ & x_1, x_2 \geq 0 \end{aligned}$$

will be solved again using the modified pivot selection procedure.

The simplex tableau is :

First x_k for which $c_k < 0$

$-z$	x_1	x_2	x_3	x_4	RHS	
1	-1	-2	0	0	0	
0	-2	2	1	0	3	
0	2	2	0	1	3	← No choice of pivot row in this case
1	0	-1	0	1/2	3/2	
0	0	4	1	1	6	← Basic variable associated with this row is x_3
0	1	1	0	1/2	3/2	← Basic variable associated with this row is x_1
1	1	0	0	1	3	
0	-4	0	1	-1	0	
0	1	1	0	1/2	3/2	

It can be seen that the solution is now optimal. The optimal basic feasible solution is :

$$z = -3, \; x_1 = 0 \text{ and } x_2 = 3/2$$

Notice that Bland's Rule has prevented cycling from occurring.

Summary
- When using the simplex methods :

  - A linear programming problem has a **non-unique solution** if in the optimal tableau, a non-basic variable has a zero coefficient in the z-row.

  - A linear programming problem is **infeasible** if at the end of phase 1, $w_{min} > 0$ (or if feasibility cannot be restored using the dual simplex method).

  - A linear programming problem is **unbounded** if at any stage during the solution the pivot selection rule breaks down *i.e.* if :

  $$S = \left\{ \frac{b_i}{a_{ik}} \mid a_{ik} > 0 \; ; \; 1 \le i \le m \right\} = \phi \quad \textit{i.e. the empty set}$$

  - A linear programming problem is **degenerate** if at any stage during the solution a zero appears on the right-hand side of a constraint row. A possible consequence of degeneracy is cycling. Cycling can usually be prevented using Bland's Rule.

3.6 Exercises 3

1. Use the simplex method to find <u>all</u> solutions of the following linear programming problems :

  (i) Maximise $z = x_1 + x_2$
       Subject to : $x_2 \le 8$
                     $-x_1 + x_2 \ge -4$
                     $x_1 + x_2 \le 12$
                     $x_1, x_2 \ge 0$

  (ii) Minimise $z = -2x_1 - 4x_2 + 4$
        Subject to : $x_2 \le 4$
                      $x_1 + 2x_2 \le 8$
                      $-2x_1 + x_2 \ge -2$
                      $x_1, x_2 \ge 0$

  (iii) Minimise $z = -x_1 - 2x_2 - 5$
         Subject to : $x_2 \le 4$
                       $x_1 + 2x_2 \le 8$
                       $-x_1 + x_2 \le 4$
                       $x_1, x_2 \ge 0$

2. Explain why a convex combination of two optimal basic feasible solutions of a linear programming problem is :

   (i)   A solution of the system of constraints.
   (ii) A feasible solution.
   (iii) An optimal solution.

3. Use the two-phase simplex method to solve the following linear programming problem :

$$\text{Minimise} \quad z = -x_1 - 2x_2 - x_3$$
$$\text{Subject to :} \quad -x_1 + 2x_2 + x_3 = 4$$
$$-2x_1 - x_3 \geq 3$$
$$3x_1 + 2x_2 - x_3 \leq 8$$
$$x_1, x_2, x_3 \geq 0$$

4. Use the simplex method to solve the following linear programming problems :

   (i) $\text{Minimise} \quad z = -6x_1 + 2x_2 + x_3 + 8$
         $\text{Subject to :} \quad x_1 - x_2 + 3x_3 \leq 6$
         $$2x_1 - x_2 + x_3 \leq 4$$
         $$x_1, x_2, x_3 \geq 0$$

   (ii) $\text{Maximise} \quad z = 4x_1 - 2x_2 - 3$
         $\text{Subject to :} \quad -2x_1 + x_2 + x_3 \leq 10$
         $$x_1 - x_2 + 2x_3 \leq 8$$
         $$x_1, x_2, x_3 \geq 0$$

5. By looking at your solutions to questions 1. and 4., state whether or not each of the linear programming problems is degenerate.

6. Consider the following linear programming problem :

$$\text{Minimise} \quad z = -10x_1 + 57x_2 + 9x_3 + 24x_4$$
$$\text{Subject to :} \quad x_1 - 11x_2 - 5x_3 + 18x_4 \leq 0$$
$$x_1 - 3x_2 - x_3 + 2x_4 \leq 0$$
$$x_1 \leq 1$$
$$x_1, x_2, x_3, x_4 \geq 0$$

   (i)   Apply the simplex algorithm to this problem. Ignore the usual rule for selecting the pivot variable but select the pivot row in the usual way. Introduce the variables into the basis in the order $x_1, x_2, x_3, x_4, x_5, x_6$ where x_5 and x_6 are the slack variables associated with the first and second constraints respectively.

   (ii) Compare the first and last tableaux in the solution to 6(i). What do you notice ? What is happening in this case ?

7. Solve the linear programming problem from question 6. using the simplex method with Bland's rule.

4. The Revised Simplex Method

4.1 Introduction

The **revised simplex method** is a matrix based algorithm that incorporates the main ideas from the simplex methods described in Chapter 1 and Chapter 2. The method is sometimes more computationally efficient than the simplex methods and is therefore commonly used for solving linear programming problems in business and industry.

In the descriptions that follow in this chapter the subscript/superscript i is used to label the variables and matrices as they are updated within the algorithms. For example, if $x^{(i)}$ denotes the current value of a variable x then $x^{(i+1)}$ denotes the next value of x. Similarly, if A_i denotes the current matrix A then A_{i+1} denotes the next matrix A, etc.

4.2 Development of the Basic Algorithm

To develop the basic revised simplex algorithm problems that contain **type 1** (*i.e.* $\leq$) constraints only will be considered. Problems of this kind can be solved using **phase 2** of the procedure. The two-phase method *i.e.* the procedure used for solving general linear programming problems will be developed later.

Consider the following general linear programming problem :

$$\text{Minimise} \quad z = c_0 + \underline{c}\,\underline{x}$$
$$\text{Subject to :} \quad A\underline{x} \leq \underline{b}$$
$$\underline{x} \geq \underline{0}$$

where $\underline{b} \geq \underline{0}$, $\underline{c}$ is a row vector, A is a matrix, $\underline{x}$ is a column vector and $\underline{b}$ is a column vector. By rearranging the objective function and adding slack variables this problem can be written in canonical form as :

$$\text{Minimise} \quad -z + \underline{c}\,\underline{x} = -c_0$$
$$\text{Subject to :} \quad A\underline{x} = \underline{b}$$
$$\underline{x} \geq \underline{0}$$

The vector $\underline{x}$ can be partitioned into :

$$\underline{x} = \begin{bmatrix} \underline{x}_B \\ .. \\ \underline{x}_N \end{bmatrix}$$

where :

- The vector $\underline{x}_B$ contains the basic variables. As in the simplex methods it is usual to treat the **slack** variables as **basic** initially. Initially, these variables are listed in the order in which they occur in the original problem. However, as the method proceeds the order of the variables in this vector is determined by the algorithm. The vector $\underline{x}_B$ is called the **current basis**. It is important to keep a record of the current basis at each stage of the solution.

- The vector $\underline{x}_N$ contains the non-basic variables. As in the simplex methods it is usual to treat the **structural** variables as **non-basic** initially. As these variables are all zero the order in which they appear in this vector is unimportant. However, it is usual to list them in ascending subscript order.

The vector $\underline{c}$ can be partitioned into :

$$\underline{c} = \left[\underline{c}_B : \underline{c}_N \right]$$

where :

- The vector $\underline{c}_B$ contains the coefficients of the basic variables in the objective function. Initially, $\underline{c}_B = \underline{0}$ _i.e._ the zero vector.

- The vector $\underline{c}_N$ contains the coefficients of the non-basic variables in the objective function.

The matrix A can be partitioned into :

$$A = \left[B : N \right]$$

where :

- The matrix B contains the columns of the matrix A associated with the **basic** variables (in the order in which they occur in the matrix A). The matrix B is called the **basis matrix**. Initially, $B = I$ _i.e._ the identity matrix.

- The matrix N contains the columns of the matrix A associated with the **non-basic** variables (in the order in which they occur in the matrix A).

Using this partition the general linear programming problem can be written as :

$$\text{Minimise} \quad -z + \left[\underline{c}_B : \underline{c}_N \right] \begin{bmatrix} \underline{x}_B \\ .. \\ \underline{x}_N \end{bmatrix} = -c_0$$

$$\text{Subject to :} \quad \left[B : N \right] \begin{bmatrix} \underline{x}_B \\ .. \\ \underline{x}_N \end{bmatrix} = \underline{b}$$

$$\underline{x}_B \geq \underline{0}, \ \underline{x}_N \geq \underline{0}$$

_i.e._ Minimise $\quad -z + \underline{c}_B \underline{x}_B + \underline{c}_N \underline{x}_N = -c_0$

Subject to : $\quad B\underline{x}_B + N\underline{x}_N = \underline{b}$

$$\underline{x}_B \geq \underline{0}, \ \underline{x}_N \geq \underline{0}$$

or in matrix form :

$$\begin{bmatrix} 1 & \underline{c}_B & \underline{c}_N \\ 0 & B & N \end{bmatrix}\begin{bmatrix} -z \\ \underline{x}_B \\ \underline{x}_N \end{bmatrix} = \begin{bmatrix} -c_0 \\ \underline{b} \end{bmatrix}$$

To solve the linear programming problem this system of linear equations must be solved for the basic variables. It can be shown that the solution can be obtained by premultiplying each side of this system by a matrix in the form :

$$T = \begin{bmatrix} 1 & -\underline{c}_B B^{-1} \\ \underline{0} & B^{-1} \end{bmatrix}$$

The matrix T is called the **transformation matrix**. Using this, the matrix representation of the simplex tableau **at any stage during phase 2** is :

$$T\begin{bmatrix} 1 & \underline{c}_B & \underline{c}_N \\ 0 & B & N \end{bmatrix}\begin{bmatrix} -z \\ \underline{x}_B \\ \underline{x}_N \end{bmatrix} = T\begin{bmatrix} -c_0 \\ \underline{b} \end{bmatrix} \quad \text{---- (1)}$$

$$\begin{bmatrix} 1 & -\underline{c}_B B^{-1} \\ \underline{0} & B^{-1} \end{bmatrix}\begin{bmatrix} 1 & \underline{c}_B & \underline{c}_N \\ 0 & B & N \end{bmatrix}\begin{bmatrix} -z \\ \underline{x}_B \\ \underline{x}_N \end{bmatrix} = \begin{bmatrix} 1 & -\underline{c}_B B^{-1} \\ \underline{0} & B^{-1} \end{bmatrix}\begin{bmatrix} -c_0 \\ \underline{b} \end{bmatrix}$$

i.e. $$\begin{bmatrix} 1 & \underline{0} & \underline{c}_N - \underline{c}_B B^{-1} N \\ \underline{0} & I & B^{-1} N \end{bmatrix}\begin{bmatrix} -z \\ \underline{x}_B \\ \underline{x}_N \end{bmatrix} = \begin{bmatrix} -c_0 - \underline{c}_B B^{-1}\underline{b} \\ B^{-1}\underline{b} \end{bmatrix} \quad \text{---- (2)}$$

Expression (1) suggests an alternative way of implementing the simplex methods *i.e.* instead of transforming one tableau to the next simply hold the original problem (in matrix form) together with the current T matrix. Once optimality is detected the optimal solution can be calculated by premultiplying the original problem (in matrix form) by the optimal T matrix. This alternative procedure is called the revised simplex method.

Calculating the T Matrix

One way of calculating the T matrix at each iteration is to write down the current basis matrix B_i, calculate B_i^{-1} and then evaluate T_i using :

$$T_i = \begin{bmatrix} 1 & -\underline{c}_B^{(i)} B_i^{-1} \\ \underline{0} & B_i^{-1} \end{bmatrix}$$

However, calculating the matrix B_i^{-1} is expensive computationally and so this approach is impractical.

Using expression (1) the ith simplex tableau can be written as :

$$
T_i \begin{bmatrix} 1 & \underline{c}_B & \underline{c}_N \\ 0 & B & N \end{bmatrix} \begin{bmatrix} -z \\ \underline{x}_B \\ \underline{x}_N \end{bmatrix} = T_i \begin{bmatrix} -c_0 \\ \underline{b} \end{bmatrix} \quad \text{---- (3)}
$$

and the $(i+1)$th simplex tableau can be written as :

$$
T_{i+1} \begin{bmatrix} 1 & \underline{c}_B & \underline{c}_N \\ 0 & B & N \end{bmatrix} \begin{bmatrix} -z \\ \underline{x}_B \\ \underline{x}_N \end{bmatrix} = T_{i+1} \begin{bmatrix} -c_0 \\ \underline{b} \end{bmatrix} \quad \text{---- (4)}
$$

In the simplex method the new tableau is obtained from the previous one by performing Jordan elimination. It can be shown that this is equivalent to premultiplying each side of the matrix representation of the previous tableau by a matrix in the form :

$$
J_i = \begin{bmatrix}
1 & 0 & . & . & 0 & \dfrac{-c_k^{(i)}}{a_{pk}^{(i)}} & 0 & . & . & 0 \\
0 & . & . & . & 0 & \dfrac{-a_{1k}^{(i)}}{a_{pk}^{(i)}} & 0 & . & . & 0 \\
. & & . & & & . & & & & . \\
. & & & . & & . & & & & . \\
. & & & & 1 & . & & & & . \\
0 & . & . & . & 0 & \dfrac{1}{a_{pk}^{(i)}} & 0 & . & . & 0 \\
. & & & & & . & 1 & & & . \\
. & & & & & . & & . & & . \\
. & & & & & . & & & . & 0 \\
0 & 0 & . & . & . & \dfrac{-a_{mk}^{(i)}}{a_{pk}^{(i)}} & 0 & . & . & 1
\end{bmatrix}
$$

The matrix J_i is called the **Jordan** matrix. This is an identity matrix that has one of its columns replaced by the **pivot column multipliers**. The column replaced is the column that corresponds to the pivot row (*e.g.* if the pivot row is row j then column j in the identity matrix is replaced by the pivot column multipliers). A justification of the formulae used in this matrix is given later.

Using the J matrix expression (4) can be written as :

$$
J_i T_i \begin{bmatrix} 1 & \underline{c}_B & \underline{c}_N \\ 0 & B & N \end{bmatrix} \begin{bmatrix} -z \\ \underline{x}_B \\ \underline{x}_N \end{bmatrix} = J_i T_i \begin{bmatrix} -c_0 \\ \underline{b} \end{bmatrix} \quad \text{---- (5)}
$$

By comparing expressions (4) and (5) it can be seen that :

$$
T_{i+1} = J_i T_i \quad \text{---- (6)}
$$

Hence, to calculate the T matrix at each iteration perform Jordan elimination on the previous one *i.e.* form the matrix J_i and then evaluate expression (6).

Calculating the J Matrix

To calculate the J matrix at each iteration it is necessary to find the **pivot**. From the pivot selection rule used in the simplex methods *i.e.*

$$\frac{b_p}{a_{pk}} = \min\left\{\frac{b_i}{a_{ik}} \mid a_{ik} > 0 \; ; \; 1 \le i \le m\right\}$$

it can be seen that the pivot calculation requires (i) the pivot variable (ii) the pivot column and (iii) the right-hand side vector.

The Pivot Variable

To find the pivot variable calculate the coefficients of the non-basic variables in the objective function (*i.e.* the elements of the vector $\underline{c}_N^{(i)}$) until a negative one is found. These elements are calculated by finding the inner product of the **first row** of the current T matrix with each of the columns from the <u>original problem</u> associated with <u>current</u> non-basic variables (in the order in which they appear in the vector $\underline{x}_N$). If all of the elements of the vector $\underline{c}_N^{(i)}$ are greater than or equal to zero then the current basic feasible solution is optimal and the algorithm **stops**.

The Pivot Column

Suppose that the first negative element of the vector $\underline{c}_N^{(i)}$ is $c_k^{(i)}$. To find the pivot column premultiply the $\underline{x}_k$ column from the <u>original problem</u> by the current T matrix.

The Right-Hand Side Vector

At the first iteration :

$$\underline{r}_0 = \begin{bmatrix} -c_0 \\ \underline{b} \end{bmatrix}$$

At the second and subsequent iterations the vector $\underline{r}_i$ is found by performing Jordan elimination on the previous one *i.e.* by calculating :

$$\underline{r}_{i+1} = J_i \underline{r}_i$$

where J_i is the J matrix from the <u>previous</u> iteration.

The Formulae for the Pivot Column Multipliers

Suppose that the pivot column is column k and that the pivot row is row p *i.e.*

$$
\begin{array}{ll}
z \text{ row} & c_k^{(i)} \\[4pt]
\text{row 1} & a_{1k}^{(i)} \\[4pt]
\cdot & \cdot \\[2pt]
\cdot & \cdot \\[2pt]
\cdot & \cdot \\[4pt]
\text{row } p & a_{pk}^{(i)} \quad \longleftarrow \quad \text{The Pivot} \\[4pt]
\cdot & \cdot \\[2pt]
\cdot & \cdot \\[2pt]
\cdot & \cdot \\[4pt]
\text{row } m & a_{mk}^{(i)}
\end{array}
$$

- To make the pivot into a <u>one</u> divide through the pivot row by the pivot *i.e.* by $a_{pk}^{(i)}$. Hence, the pivot column multiplier for the pivot row is $\dfrac{1}{a_{pk}^{(i)}}$.

Assuming that the pivot is now a <u>one</u> :

- To put a <u>zero</u> in the z row subtract $c_k^{(i)}$ times the pivot row from the z row. Hence, the pivot column multiplier for the z row is $-c_k^{(i)}$ times the pivot column multiplier for the pivot row. That is : $-c_k^{(i)} \times \dfrac{1}{a_{pk}^{(i)}}$ *i.e.* $\dfrac{-c_k^{(i)}}{a_{pk}^{(i)}}$.

- To put a <u>zero</u> in row 1 subtract $a_{1k}^{(i)}$ times the pivot row from row 1. Hence, the pivot column multiplier for row 1 is $-a_{1k}^{(i)}$ times the pivot column multiplier for the pivot row. That is :

$$
-a_{1k}^{(i)} \times \frac{1}{a_{pk}^{(i)}} \quad \textit{i.e.} \quad \frac{-a_{1k}^{(i)}}{a_{pk}^{(i)}} \, .
$$

Continuing in this way :

- To put a <u>zero</u> in row m subtract $a_{mk}^{(i)}$ times the pivot row from row m. Hence, the pivot column multiplier for row m is $-a_{mk}^{(i)}$ times the pivot column multiplier for the pivot row. That is :

$$
-a_{mk}^{(i)} \times \frac{1}{a_{pk}^{(i)}} \quad \textit{i.e.} \quad \frac{-a_{mk}^{(i)}}{a_{pk}^{(i)}} \, .
$$

4.3 Summary of the Basic Algorithm
The basic revised simplex algorithm can be summarised as follows :

Step 1
Write the linear programming problem in **canonical form** *i.e.*

$$\text{Minimise} \quad -z + \underline{c}\,\underline{x} = -c_0$$
$$\text{Subject to :} \quad A\underline{x} = \underline{b}$$
$$\underline{x} \geq \underline{0}$$

Step 2
Partition the vector $\underline{x}$ into :

$$\underline{x} = \begin{bmatrix} \underline{x}_B \\ .. \\ \underline{x}_N \end{bmatrix}$$

Partition the vector $\underline{c}$ into :

$$\underline{c} = \begin{bmatrix} \underline{c}_B : \underline{c}_N \end{bmatrix}$$

Partition the matrix A into :

$$A = \begin{bmatrix} B : N \end{bmatrix}$$

Step 3
Calculate the initial T matrix using :

$$T_0 = \begin{bmatrix} 1 & -\underline{c}_B B^{-1} \\ \underline{0} & B^{-1} \end{bmatrix}$$

Step 4
Calculate the J matrix *i.e.*

- **Find the pivot variable.** Calculate the elements of the vector $\underline{c}_N^{(i)}$ (until a negative one is found). Find the inner product of the **first row** of the current T matrix with the columns from the original problem associated with current non-basic variables (in the order in which they appear in the vector $\underline{x}_N$).

  If all of the elements of the vector $\underline{c}_N^{(i)}$ are greater than or equal to zero, the current basic feasible solution is **optimal**. In this case stop and go to **Step 7**.

  If $c_k^{(i)} < 0$ then x_k is the pivot variable and x_k joins the basis.

- **Find the pivot column.** Find the product of the current T matrix with the $\underline{x}_k$ column from the original problem.

- **Find the right-hand side vector.**
  At the first iteration :

$$\underline{r}_0 = \begin{bmatrix} -c_0 \\ \underline{b} \end{bmatrix}$$

At the second and subsequent iterations :

$$\underline{r}_{i+1} = J_i \underline{r}_i$$

where J_i is the J matrix from the <u>previous</u> iteration.

- **Find the pivot.** Use the pivot selection rule :

$$\frac{b_p}{a_{pk}} = \min\left\{\frac{b_i}{a_{ik}} \ \middle|\ a_{ik} > 0 \ ; \ 1 \le i \le m\right\}$$

- **Update the current basis.** Annotate the pivot column with the current basis remembering that the first row of this vector is associated with $(-z)$. If the variable in the pivot row is x_j then x_j leaves the basis.

- **Find the pivot column multipliers.** Use the formulae :

$$\frac{-c_k^{(i)}}{a_{pk}^{(i)}}$$

$$\frac{-a_{1k}^{(i)}}{a_{pk}^{(i)}}$$

$$\vdots$$

$$\frac{1}{a_{pk}^{(i)}}$$

$$\vdots$$

$$\frac{-a_{mk}^{(i)}}{a_{pk}^{(i)}}$$

- **Form the J matrix.**

Step 5
Calculate the T matrix. Use the formula :

$$T_{i+1} = J_i T_i$$

where J_i is the J matrix calculated in **Step 4**.

Step 6
Go back to **Step 4**.

Step 7
Extract the optimal solution *i.e.*

- Calculate the optimal right-hand side vector. Use the formula :

$$\underline{r}_{i+1} = J_i\,\underline{r}_i$$

  where J_i is the J matrix calculated in **Step 4**.

- Annotate the vector $\underline{r}_{i+1}$ with the current basis remembering that the first row of this vector is associated with $(-z)$.

- Read off the optimal values of z and the basic variables. The value of each non-basic variable is zero.

Example 4.1
Use the revised simplex method to solve the following linear programming problem :

$$\begin{aligned}
\text{Maximise} \quad & z = 5x_1 + 3x_2 \\
\text{Subject to :} \quad & 4x_1 + 5x_2 \le 1000 \\
& 5x_1 + 2x_2 \le 1000 \\
& 3x_1 + 8x_2 \le 1200 \\
& x_1, x_2 \ge 0
\end{aligned}$$

Step 1
This problem can be written in canonical form as :

$$\begin{aligned}
\text{Minimise} \quad & -\bar{z} - 5x_1 - 3x_2 = 0 \\
\text{Subject to :} \quad & 4x_1 + 5x_2 + x_3 = 1000 \\
& 5x_1 + 2x_2 + x_4 = 1000 \\
& 3x_1 + 8x_2 + x_5 = 1200 \\
& x_1, x_2, x_3, x_4, x_5 \ge 0
\end{aligned}$$

Step 2
Partitioning this problem :

$$\underline{x}_B = \begin{bmatrix} x_3 \\ x_4 \\ x_5 \end{bmatrix} \quad , \quad \underline{x}_N = \begin{bmatrix} x_1 \\ x_2 \end{bmatrix}$$

$$\underline{c}_B = \begin{bmatrix} 0 & 0 & 0 \end{bmatrix} \quad , \quad \underline{c}_N = \begin{bmatrix} -5 & -3 \end{bmatrix}$$

$$B = \begin{bmatrix} 1 & 0 & 0 \\ 0 & 1 & 0 \\ 0 & 0 & 1 \end{bmatrix} \quad , \quad N = \begin{bmatrix} 4 & 5 \\ 5 & 2 \\ 3 & 8 \end{bmatrix}$$

Step 3

The initial T matrix is given by :

$$T_0 = \begin{bmatrix} 1 & -\underline{c}_B B^{-1} \\ \underline{0} & B^{-1} \end{bmatrix}$$

Here :

$$B^{-1} = \begin{bmatrix} 1 & 0 & 0 \\ 0 & 1 & 0 \\ 0 & 0 & 1 \end{bmatrix} \quad , \quad -\underline{c}_B B^{-1} = -\begin{bmatrix} 0 & 0 & 0 \end{bmatrix} \begin{bmatrix} 1 & 0 & 0 \\ 0 & 1 & 0 \\ 0 & 0 & 1 \end{bmatrix} = \begin{bmatrix} 0 & 0 & 0 \end{bmatrix}$$

Hence :

$$T_0 = \begin{bmatrix} 1 & 0 & 0 & 0 \\ 0 & 1 & 0 & 0 \\ 0 & 0 & 1 & 0 \\ 0 & 0 & 0 & 1 \end{bmatrix}$$

Iteration 1

Step 4

Calculating the J matrix :

- **The pivot variable :**

First row of current T matrix

$$c_1^{(1)} = \begin{bmatrix} 1 & 0 & 0 & 0 \end{bmatrix} \begin{bmatrix} -5 \\ 4 \\ 5 \\ 3 \end{bmatrix} = -5 \quad i.e. \ c_1^{(1)} < 0 \ \text{Hence } x_1 \text{ is the pivot variable } i.e. \ x_1 \text{ joins the basis.}$$

Coefficients of the first non-basic variable x_1

- **The pivot column :**
  The pivot column is :

$$\underline{x}_1^{(1)}=T_0\underline{x}_1^{(0)} \quad i.e. \quad \underline{x}_1^{(1)}=\begin{bmatrix}1&0&0&0\\0&1&0&0\\0&0&1&0\\0&0&0&1\end{bmatrix}\begin{bmatrix}-5\\4\\5\\3\end{bmatrix}=\begin{bmatrix}-5\\4\\5\\3\end{bmatrix}$$

- **The right-hand side vector :**
  At this iteration :

$$\underline{r}_0=\begin{bmatrix}-c_0\\\underline{b}\end{bmatrix} \quad i.e. \ \underline{r}_0=\begin{bmatrix}0\\1000\\1000\\1200\end{bmatrix}$$

- **The pivot :**
  Using the pivot selection rule :

$$S=\{1000/4,1000/5,1200/3\}$$

  The smallest ratio is $1000/5$. Hence, the pivot is the 5 in row 3 of the pivot column.

- **Updating the current basis :**
  Annotating the pivot column with the current basis :

$$\begin{matrix}(-\bar{z})\\x_3\\x_4\\x_5\end{matrix}\begin{bmatrix}-5\\4\\5\\3\end{bmatrix}\longleftarrow \text{Pivot}$$

  i.e. x_4 leaves the basis. Hence, at the next iteration :

$$\underline{x}_B^{(1)}=\begin{bmatrix}x_3\\x_1\\x_5\end{bmatrix} \quad , \quad \underline{x}_N^{(1)}=\begin{bmatrix}x_2\\x_4\end{bmatrix}$$

- **The pivot column multipliers :**
  The pivot column multipliers are :

$$-(-5)\text{x}1/5=1$$
$$-4\text{x}1/5=-4/5$$
$$1/5$$
$$-3\text{x}1/5=-3/5$$

Hence, the J matrix is :

$$J_0 = \begin{bmatrix} 1 & 0 & 1 & 0 \\ 0 & 1 & -4/5 & 0 \\ 0 & 0 & 1/5 & 0 \\ 0 & 0 & -3/5 & 1 \end{bmatrix}$$

Step 5
Calculating the T matrix :

$$T_{i+1} = J_i T_i$$

$$T_1 = J_0 T_0$$

$$i.e. \quad T_1 = \begin{bmatrix} 1 & 0 & 1 & 0 \\ 0 & 1 & -4/5 & 0 \\ 0 & 0 & 1/5 & 0 \\ 0 & 0 & -3/5 & 1 \end{bmatrix} \begin{bmatrix} 1 & 0 & 0 & 0 \\ 0 & 1 & 0 & 0 \\ 0 & 0 & 1 & 0 \\ 0 & 0 & 0 & 1 \end{bmatrix} = \begin{bmatrix} 1 & 0 & 1 & 0 \\ 0 & 1 & -4/5 & 0 \\ 0 & 0 & 1/5 & 0 \\ 0 & 0 & -3/5 & 1 \end{bmatrix}$$

Step 6
Go back to **Step 4**.

Iteration 2

Step 4
Calculating the J matrix :

- **The pivot variable :**

$$c_2^{(2)} = \begin{bmatrix} 1 & 0 & 1 & 0 \end{bmatrix} \begin{bmatrix} -3 \\ 5 \\ 2 \\ 8 \end{bmatrix} = -1 \quad i.e. \ c_2^{(2)} < 0. \text{ Hence } x_2 \text{ is the pivot variable } i.e. \ x_2 \text{ joins the basis.}$$

- **The pivot column :**
  The pivot column is :

$$\underline{x}_2^{(2)} = T_1 \underline{x}_2^{(0)} \quad i.e. \quad \underline{x}_2^{(2)} = \begin{bmatrix} 1 & 0 & 1 & 0 \\ 0 & 1 & -4/5 & 0 \\ 0 & 0 & 1/5 & 0 \\ 0 & 0 & -3/5 & 1 \end{bmatrix} \begin{bmatrix} -3 \\ 5 \\ 2 \\ 8 \end{bmatrix} = \begin{bmatrix} -1 \\ 17/5 \\ 2/5 \\ 34/5 \end{bmatrix}$$

- **The right-hand side vector :**

At this iteration :

$$\underline{r}_1 = J_0 \underline{r}_0 \quad i.e. \ \underline{r}_1 = \begin{bmatrix} 1 & 0 & 1 & 0 \\ 0 & 1 & -4/5 & 0 \\ 0 & 0 & 1/5 & 0 \\ 0 & 0 & -3/5 & 1 \end{bmatrix} \begin{bmatrix} 0 \\ 1000 \\ 1000 \\ 1200 \end{bmatrix} = \begin{bmatrix} 1000 \\ 200 \\ 200 \\ 600 \end{bmatrix}$$

- **The pivot :**
Using the pivot selection rule :

$$S = \left\{ \frac{200}{17/5}, \frac{200}{2/5}, \frac{600}{34/5} \right\}$$

The smallest ratio is $\dfrac{200}{17/5}$. Hence, the pivot is the $17/5$ in row 2 of the pivot column.

- **Updating the current basis :**
Annotating the pivot column with the current basis :

$$\begin{array}{c} (-\bar{z}) \\ x_3 \\ x_1 \\ x_5 \end{array} \begin{bmatrix} -1 \\ 17/5 \\ 2/5 \\ 34/5 \end{bmatrix} \longleftarrow \text{Pivot}$$

i.e. x_3 leaves the basis. Hence, at the next iteration :

$$\underline{x}_B^{(2)} = \begin{bmatrix} x_2 \\ x_1 \\ x_5 \end{bmatrix} \quad , \quad \underline{x}_N^{(2)} = \begin{bmatrix} x_3 \\ x_4 \end{bmatrix}$$

- **The pivot column multipliers :**
The pivot column multipliers are :

$$-(-1) \times 5/17 = 5/17$$
$$5/17$$
$$-(2/5) \times 5/17 = -2/17$$
$$-(34/5) \times 5/17 = -2$$

Hence, the J matrix is :

$$J_1 = \begin{bmatrix} 1 & 5/17 & 0 & 0 \\ 0 & 5/17 & 0 & 0 \\ 0 & -2/17 & 1 & 0 \\ 0 & -2 & 0 & 1 \end{bmatrix}$$

Step 5
Calculating the T matrix :

$$T_{i+1} = J_i T_i$$

$$T_2 = J_1 T_1$$

$$i.e. \quad T_2 = \begin{bmatrix} 1 & 5/17 & 0 & 0 \\ 0 & 5/17 & 0 & 0 \\ 0 & -2/17 & 1 & 0 \\ 0 & -2 & 0 & 1 \end{bmatrix} \begin{bmatrix} 1 & 0 & 1 & 0 \\ 0 & 1 & -4/5 & 0 \\ 0 & 0 & 1/5 & 0 \\ 0 & 0 & -3/5 & 1 \end{bmatrix} = \begin{bmatrix} 1 & 5/17 & 13/17 & 0 \\ 0 & 5/17 & -4/17 & 0 \\ 0 & -2/17 & 5/17 & 0 \\ 0 & -2 & 1 & 1 \end{bmatrix}$$

Step 6
Go back to **Step 4**.

Iteration 3

Step 4
Calculating the J matrix :

- **The pivot variable :**

$$c_3^{(3)} = \begin{bmatrix} 1 & 5/17 & 13/17 & 0 \end{bmatrix} \begin{bmatrix} 0 \\ 1 \\ 0 \\ 0 \end{bmatrix} = 5/17 \quad i.e. \quad c_3^{(3)} > 0$$

$$c_4^{(3)} = \begin{bmatrix} 1 & 5/17 & 13/17 & 0 \end{bmatrix} \begin{bmatrix} 0 \\ 0 \\ 1 \\ 0 \end{bmatrix} = 13/17 \quad i.e. \quad c_4^{(3)} > 0$$

Since all of the elements of the vector $\underline{c}_N^{(3)}$ are greater than or equal to zero then the current basic feasible solution is **optimal**.

Step 7
- The optimal right-hand side vector is :

$$\underline{r}_2 = J_1 \underline{r}_1$$

$$i.e. \quad \underline{r}_2 = \begin{bmatrix} 1 & 5/17 & 0 & 0 \\ 0 & 5/17 & 0 & 0 \\ 0 & -2/17 & 1 & 0 \\ 0 & -2 & 0 & 1 \end{bmatrix} \begin{bmatrix} 1000 \\ 200 \\ 200 \\ 600 \end{bmatrix} = \begin{bmatrix} 18000/17 \\ 1000/17 \\ 3000/17 \\ 200 \end{bmatrix}$$

- Annotating this vector with the current basis :

$$
\begin{array}{c}
(-\bar{z}) \\
x_2 \\
x_1 \\
x_5
\end{array}
\begin{bmatrix}
18000/17 \\
1000/17 \\
3000/17 \\
200
\end{bmatrix}
$$

Hence, the optimal solution of the linear programming problem is :

$$-\bar{z} = 18000/17 \ i.e. \ z = 18000/17, \ x_2 = 1000/17, \ x_1 = 3000/17, \ x_5 = 200, \ x_3 = 0 \ \text{and} \ x_4 = 0$$

4.4 The Tableau Version of the Basic Algorithm

The revised simplex method can be implemented using a tableau. This version of the algorithm is the most convenient to use when performing the calculations by hand rather than using a computer. A commonly used form of this tableau is shown below.

	-z/-z̄ + Original Basic Variables	Variables	PC	RHS	Comments
	Transformation Matrix	Variable Coefficients			

-z/-z̄ + Current Basic Variables

Pivot Column

Right-hand Side Vector

When using this version of the revised simplex method it is not necessary to calculate and use the J matrix. Jordan elimination can be applied to the appropriate elements within the tableau in the same way as it is in the other simplex methods.

The Procedure

To solve a linear programming problem using the tableau version of the revised simplex method :

1. Write the problem in canonical form, partition it, calculate the initial T matrix using the formula :

$$T_0 = \begin{bmatrix} 1 & -\underline{c}_B B^{-1} \\ \underline{0} & B^{-1} \end{bmatrix}$$

write down the initial tableau and initialise the label i *i.e.* let $i = 1$

2. Find the pivot variable. To do this calculate the coefficients of the non-basic variables in the z-row (*i.e.* the elements of the vector $\underline{c}_N^{(i)}$) until a negative one is found. These values are calculated by finding the inner product of the **first row** of the current T matrix with each of the columns from the **original problem** associated with **current** non-basic variables (in the order in which they appear in the vector $\underline{x}_N$). If the first negative coefficient is c_k then x_k is the pivot variable. If all of the elements of the vector $\underline{c}_N^{(i)}$ are greater than or equal to zero then the current basic feasible solution is optimal and the algorithm **stops**. Otherwise :

3. Calculate the pivot column. To do this multiply the column associated with the pivot variable *i.e.* x_k by the current T matrix.

4. Find the pivot row. This is done in the same way as it is in the ordinary simplex and two-phase simplex methods *i.e.* by dividing the elements in the right-hand side vector by the positive coefficients in the pivot column and choosing the smallest ratio. Suppose that the pivot row is the row associated with the variable x_j in the tableau.

5. Update the current basis labels to indicate that x_k is joining the basis and x_j is leaving the basis.

6. Perform Jordan elimination to transform the pivot column into a column of the identity matrix. In the tableau version of the revised simplex method the row operations are performed on the right-hand side vector and the current T matrix **only**.

7. Let $i = i + 1$ and go back to Step 2.

When optimality has been achieved, the optimal solution can be read off from the final tableau by reading across the tableau from left to right *i.e.* from the first column to the last column.

Example 4.2

Use the revised simplex method to solve the following linear programming problem :

$$\begin{aligned} \text{Minimise} \quad & z = 2x_1 - 2x_2 - x_3 + 2 \\ \text{Subject to :} \quad & -2x_1 + x_2 + x_3 \leq 4 \\ & 3x_1 - x_2 + 2x_3 \leq 2 \\ & x_1, x_2, x_3 \geq 0 \end{aligned}$$

Solution

Step 1 : Write the problem in canonical form, partition it, calculate the initial T matrix, write down the initial tableau and initialise the label i

This problem can be written in canonical form as :

Minimise $\quad -z + 2x_1 - 2x_2 - x_3 = -2$

Subject to : $\quad -2x_1 + x_2 + x_3 + x_4 = 4$

$\quad\quad\quad\quad 3x_1 - x_2 + 2x_3 + x_5 = 2$

$\quad\quad\quad\quad x_1, x_2, x_3, x_4, x_5 \geq 0$

Partitioning the problem :

$$\underline{x}_B = \begin{bmatrix} x_4 \\ x_5 \end{bmatrix} \quad , \quad \underline{x}_N = \begin{bmatrix} x_1 \\ x_2 \\ x_3 \end{bmatrix}$$

$$\underline{c}_B = \begin{bmatrix} 0 & 0 \end{bmatrix} \quad , \quad \underline{c}_N = \begin{bmatrix} 2 & -2 & -1 \end{bmatrix}$$

$$B = \begin{bmatrix} 1 & 0 \\ 0 & 1 \end{bmatrix} \quad , \quad N = \begin{bmatrix} -2 & 1 & 1 \\ 3 & -1 & 2 \end{bmatrix}$$

The initial T matrix is given by the formula :

$$T_0 = \begin{bmatrix} 1 & -\underline{c}_B B^{-1} \\ \underline{0} & B^{-1} \end{bmatrix}$$

Here :

$$B^{-1} = \begin{bmatrix} 1 & 0 \\ 0 & 1 \end{bmatrix} \quad , \quad -\underline{c}_B B^{-1} = -\begin{bmatrix} 0 & 0 \end{bmatrix} \begin{bmatrix} 1 & 0 \\ 0 & 1 \end{bmatrix} = \begin{bmatrix} 0 & 0 \end{bmatrix}$$

Hence :

$$T_0 = \begin{bmatrix} 1 & 0 & 0 \\ 0 & 1 & 0 \\ 0 & 0 & 1 \end{bmatrix}$$

The initial revised simplex tableau is :

	$-z$	x_4	x_5	x_1	x_2	x_3	x_4	x_5	PC	RHS
$-z$	1	0	0	2	-2	-1	0	0		-2
x_4	0	1	0	-2	1	1	1	0		4
x_5	0	0	1	3	-1	2	0	1		2

Let $i = 1$

Step 2 : Find the pivot variable :

$$c_1^{(i)} = c_1^{(1)} = \underline{T}_1 \underline{x}_1 = 2 \quad c_2^{(i)} = c_2^{(1)} = \underline{T}_1 \underline{x}_2 = -2$$

First row of T matrix
$\underline{T}_1$

Since $c_2^{(1)} < 0$, x_2 is the pivot variable *i.e.* x_2 will join the basis

	$-z$	x_4	x_5	x_1	x_2	x_3	x_4	x_5	PC	RHS
$-z$	1	0	0	2	-2	-1	0	0		-2
x_4	0	1	0	-2	1	1	1	0		4
x_5	0	0	1	3	-1	2	0	1		2

$\underline{x}_1 \quad \underline{x}_2 \quad \underline{x}_3 \quad \underline{x}_4 \quad \underline{x}_5$

Step 3 : Calculate the pivot column :

$T\underline{x}_2$

	$-z$	x_4	x_5	x_1	x_2	x_3	x_4	x_5	PC	RHS
$-z$	1	0	0	2	-2	-1	0	0	-2	-2
x_4	0	1	0	-2	1	1	1	0	1	4
x_5	0	0	1	3	-1	2	0	1	-1	2

Step 4 : Find the pivot row :

	$-z$	x_4	x_5	x_1	x_2	x_3	x_4	x_5	PC	RHS	Ratio
$-z$	1	0	0	2	-2	-1	0	0	-2	-2	N/A
x_4	0	1	0	-2	1	1	1	0	1	4	$4/1 = 4$
x_5	0	0	1	3	-1	2	0	1	-1	2	N/A

x_4 will leave the basis

Pivot

Smallest Ratio

Step 5 : Update the current basis labels :

	$-z$	x_4	x_5	x_1	x_2	x_3	x_4	x_5	PC	RHS
$-z$	1	0	0	2	-2	-1	0	0	-2	-2
x_2	0	1	0	-2	1	1	1	0	1	4
x_5	0	0	1	3	-1	2	0	1	-1	2

Label updated to indicate that x_2 has
joined the basis and x_4 has left the basis

Step 6 : Perform Jordan elimination on the right-hand side vector and the current T matrix :

	$-z$	x_4	x_5	x_1	x_2	x_3	x_4	x_5	PC	RHS	Row Operations Performed
$-z$	1	2	0	2	-2	-1	0	0		6	Row 1 + 2 x Pivot Row
x_2	0	1	0	-2	1	1	1	0		4	Pivot Row
x_5	0	1	1	3	-1	2	0	1		6	Row 3 + Pivot Row

T matrix updated   Although the pivot column is now a column of the identity matrix it is left blank here because it is recalculated in **Step 3** of the next iteration   Right-hand side vector updated

Step 7 : Let $i = i+1$ *i.e.* $i = 2$ and go back to Step 2.

The complete (and annotated) revised simplex tableau for this problem is :

	$-z$	x_4	x_5	x_1	x_2	x_3	x_4	x_5	PC	RHS	Comments
$-z$	1	0	0	2	-2	-1	0	0	-2	-2	$c_1^{(1)} > 0$. $c_2^{(1)} < 0$ *i.e.* x_2 joins basis
x_4	0	1	0	-2	1	1	1	0	1	4	Smallest ratio = 4 *i.e.* x_4 leaves basis
x_5	0	0	1	3	-1	2	0	1	-1	2	
$-z$	1	2	0	2	-2	-1	0	0	-2	6	$c_1^{(2)} < 0$ *i.e.* x_1 joins basis
x_2	0	1	0	-2	1	1	1	0	-2	4	
x_5	0	1	1	3	-1	2	0	1	1	6	Smallest ratio = 6 *i.e.* x_5 leaves basis
$-z$	1	4	2	2	-2	-1	0	0		18	$c_3^{(3)} > 0$, $c_4^{(3)} > 0$, $c_5^{(3)} > 0$
x_2	0	3	2	-2	1	1	1	0		16	Solution optimal
x_1	0	1	1	3	-1	2	0	1		6	

The final tableau shows that the solution is now optimal. Reading across the tableau from left to right the optimal solution is $-z = 18$ *i.e.* $z = -18$, $x_1 = 6$, $x_2 = 16$ and $x_3 = 0$.

4.5 Exercises 4a

1. Use the revised simplex method to solve the following linear programming problem :

$$\text{Maximise} \quad z = x_1 + x_2 - 3/2$$
$$\text{Subject to :} \quad x_1 - x_2 \geq -1/2$$
$$x_1 \leq 3$$
$$x_1, x_2 \geq 0$$

2. (i) Write the initial simplex tableau for the linear programming problem in question 1. in matrix form.

   (ii) Use your optimal T matrix from question 1. and your initial simplex tableau from part (i) of this question to write down the optimal simplex tableau for the linear programming problem in question 1.

3. Show that the linear programming problem :

$$\begin{bmatrix} 1 & \underline{c}_B & \underline{c}_N \\ 0 & B & N \end{bmatrix} \begin{bmatrix} -z \\ \underline{x}_B \\ \underline{x}_N \end{bmatrix} = \begin{bmatrix} -c_0 \\ \underline{b} \end{bmatrix}$$

   can be solved for the basic variables by premultiplying each side of this expression by a transformation matrix in the form :

$$T = \begin{bmatrix} 1 & -\underline{c}_B B^{-1} \\ \underline{0} & B^{-1} \end{bmatrix}$$

   Hint : A linear programming problem can be solved by setting the non-basic variables to zero and then solving the system for the basic variables.

4. The Ravnescroft Foundry produces iron and steel for the car industry. The foundry buys two grades of ore for the production process. Grade 1 ore costs £600 per ton and grade 2 ore costs £900 per ton. Each 100 tons of grade 1 ore makes 30 tons of iron and 50 tons of steel. Each 100 tons of grade 2 ore makes 60 tons of iron and 10 tons of steel. The foundry cannot process more than 10,000 tons of ore each year. The suppliers cannot provide more than 6000 tons of grade 1 ore and 8000 tons of grade 2 ore each year. The foundry has a market for 5000 tons of iron and 3200 tons of steel each year. Iron sells for £2000 per ton and steel sells for £1000 per ton.

   (i) Formulate a linear programming model from which the manager of the foundry can determine the amount of each grade ore to purchase each year in order to maximise the total profit.

   (ii) Solve your linear programming model from 4(i) using the revised simplex method and interpret the solution you obtain.

4.6 The Two-Phase Revised Simplex Method

The basic algorithm can be extended to deal with general linear programming problems. Problems that contain a variety of constraint types must be solved using a two-phase procedure. The first stage in the process is to write the linear programming problem in canonical form. This is done in the same way as when using the two-phase simplex method *i.e.* by :

- Converting the problem to a minimisation problem (if necessary).

- Converting the constraints into equations by introducing slack, surplus and artificial variables where necessary.

- Forming a new objective function w that is the sum of the artificial variables.

- Adding the constraints that contain artificial variables.

- Subtracting this expression from the w-row.

Suppose that the canonical form of the linear programming problem is then :

$$\text{Minimise} \quad -z + \underline{c}\,\underline{x} = -c_0$$
$$-w + \underline{d}\,\underline{x} = -d_0$$
$$\text{Subject to :} \quad A\underline{x} = \underline{b}$$
$$\underline{x} \geq \underline{0}$$

where $\underline{b} \geq \underline{0}$, $\underline{c}$ is a row vector, $\underline{d}$ is a row vector, A is a matrix, $\underline{x}$ is a column vector and $\underline{b}$ is a column vector. In the **two-phase revised simplex method** the z-row is <u>always</u> the **first** row in the tableau and the w-row is <u>always</u> the **second** row. The order of these rows cannot be changed as it can in the ordinary two-phase simplex method.

Treating the **slack** and **artificial** variables as **basic** initially and the **structural** and **surplus** variables as **non-basic** initially this problem can be partitioned in the following way :

- The vector $\underline{x}$ can be partitioned into :

$$\underline{x} = \begin{bmatrix} \underline{x}_B \\ .. \\ \underline{x}_N \end{bmatrix}$$

- The vector $\underline{c}$ can be partitioned into :

$$\underline{c} = \begin{bmatrix} \underline{c}_B & : & \underline{c}_N \end{bmatrix}$$

- The vector $\underline{d}$ can be partitioned into :

$$\underline{d} = \begin{bmatrix} \underline{d}_B & : & \underline{d}_N \end{bmatrix}$$

- The matrix A can be partitioned into :

$$A = \begin{bmatrix} B & : & N \end{bmatrix}$$

Using this partition the linear programming problem can be written as :

$$\text{Minimise} \quad -z + \begin{bmatrix} \underline{c}_B : \underline{c}_N \end{bmatrix} \begin{bmatrix} \underline{x}_B \\ .. \\ \underline{x}_N \end{bmatrix} = -c_0$$

$$-w + \begin{bmatrix} \underline{d}_B : \underline{d}_N \end{bmatrix} \begin{bmatrix} \underline{x}_B \\ .. \\ \underline{x}_N \end{bmatrix} = -w_0$$

$$\text{Subject to :} \quad \begin{bmatrix} B : N \end{bmatrix} \begin{bmatrix} \underline{x}_B \\ .. \\ \underline{x}_N \end{bmatrix} = \underline{b}$$

$$\underline{x}_B \geq \underline{0}, \; \underline{x}_N \geq \underline{0}$$

i.e. Minimise

$$-z + \underline{c}_B \underline{x}_B + \underline{c}_N \underline{x}_N = -c_0$$

$$-w + \underline{d}_B \underline{x}_B + \underline{d}_N \underline{x}_N = -w_0$$

Subject to :

$$B \underline{x}_B + N \underline{x}_N = \underline{b}$$

$$\underline{x}_B \geq \underline{0}, \; \underline{x}_N \geq \underline{0}$$

or in matrix form :

$$\begin{bmatrix} 1 & 0 & \underline{c}_B & \underline{c}_N \\ 0 & 1 & \underline{d}_B & \underline{d}_N \\ 0 & 0 & B & N \end{bmatrix} \begin{bmatrix} -z \\ -w \\ \underline{x}_B \\ \underline{x}_N \end{bmatrix} = \begin{bmatrix} -c_0 \\ -w_0 \\ \underline{b} \end{bmatrix}$$

To solve the linear programming problem this system of linear equations must be solved for the basic variables. It can be shown that this solution is obtained by premultiplying each side of this expression by a matrix in the form :

$$T = \begin{bmatrix} 1 & 0 & -\underline{c}_B B^{-1} \\ 0 & 1 & -\underline{d}_B B^{-1} \\ \underline{0} & \underline{0} & B^{-1} \end{bmatrix}$$

where T is the **transformation matrix**. Using this, the matrix representation of the simplex tableau **at any stage during phase 1** is :

$$T \begin{bmatrix} 1 & 0 & \underline{c}_B & \underline{c}_N \\ 0 & 1 & \underline{d}_B & \underline{d}_N \\ 0 & 0 & B & N \end{bmatrix} \begin{bmatrix} -z \\ -w \\ \underline{x}_B \\ \underline{x}_N \end{bmatrix} = T \begin{bmatrix} -c_0 \\ -w_0 \\ \underline{b} \end{bmatrix} \quad ---- (1)$$

$$\text{i.e.} \quad \begin{bmatrix} 1 & 0 & -\underline{c}_B B^{-1} \\ 0 & 1 & -\underline{d}_B B^{-1} \\ \underline{0} & \underline{0} & B^{-1} \end{bmatrix} \begin{bmatrix} 1 & 0 & \underline{c}_B & \underline{c}_N \\ 0 & 1 & \underline{d}_B & \underline{d}_N \\ 0 & 0 & B & N \end{bmatrix} \begin{bmatrix} -z \\ -w \\ x_B \\ x_N \end{bmatrix} = \begin{bmatrix} 1 & 0 & -\underline{c}_B B^{-1} \\ 0 & 1 & -\underline{d}_B B^{-1} \\ \underline{0} & \underline{0} & B^{-1} \end{bmatrix} \begin{bmatrix} -c_0 \\ -w_0 \\ \underline{b} \end{bmatrix}$$

$$\text{i.e.} \quad \begin{bmatrix} 1 & 0 & \underline{0} & \underline{c}_N - \underline{c}_B B^{-1} N \\ 0 & 1 & \underline{0} & \underline{d}_N - \underline{d}_B B^{-1} N \\ \underline{0} & \underline{0} & I & B^{-1} N \end{bmatrix} \begin{bmatrix} -z \\ -w \\ x_B \\ x_N \end{bmatrix} = \begin{bmatrix} -c_0 - \underline{c}_B B^{-1} \underline{b} \\ -w_0 - \underline{d}_B B^{-1} \underline{b} \\ B^{-1} \underline{b} \end{bmatrix} \quad \text{---- (2)}$$

The Two-Phase Procedure
Each phase of the two-phase procedure is similar to the basic algorithm described earlier.

In **phase 1** :

- The aim is to minimise w *i.e.* to achieve **feasibility**.

- The z-row plays a purely passive role *i.e.* it is transformed by Jordan elimination but is never selected as the pivot row.

- To find the pivot variable, calculate the coefficients of the non-basic variables in the w-row (*i.e.* the elements of the vector $\underline{d}_N^{(i)}$) until a negative one is found. These elements are calculated by finding the inner product of the **second row** of the current T matrix with each of the columns from the original problem associated with current non-basic variables (in the order in which they appear in the vector $\underline{x}_N$). If all of the elements of the vector $\underline{d}_N^{(i)}$ are greater than or equal to zero then phase 1 is complete and the value of w_{min} can be examined. If $w_{min} = 0$, the linear programming problem is feasible and phase 2 can begin. If $w_{min} > 0$ the linear programming problem is infeasible and the algorithm **stops**.

  Note : During these calculations the non-basic artificial variables are ignored as these variables are never allowed to rejoin the basis.

In **phase 2** :

- The aim is to minimise z *i.e.* to achieve **optimality**.

- The w-row is treated as a constraint and $(-w)$ is treated as a non-negative variable. At the start of phase 2 $(-w)$ is treated as a basic variable. The w-row is transformed by Jordan elimination and can be selected as the pivot row.

- To find the pivot variable calculate the coefficients of the non-basic variables in the z-row (*i.e.* the elements of the vector $\underline{c}_N^{(i)}$) until a negative one is found. These elements are calculated by finding the inner product of the **first row** of the current T matrix with each of the columns from the original problem associated with current non-basic variables (in the order in which they appear in the vector $\underline{x}_N$). If all of the elements of the vector $\underline{c}_N^{(i)}$ are greater than or equal to zero then the current basic feasible solution is optimal and the algorithm **stops**.

4.7 The Tableau Version of the Two-Phase Algorithm

The two-phase revised simplex method can also be implemented using a tableau. A commonly used form of this tableau is shown below.

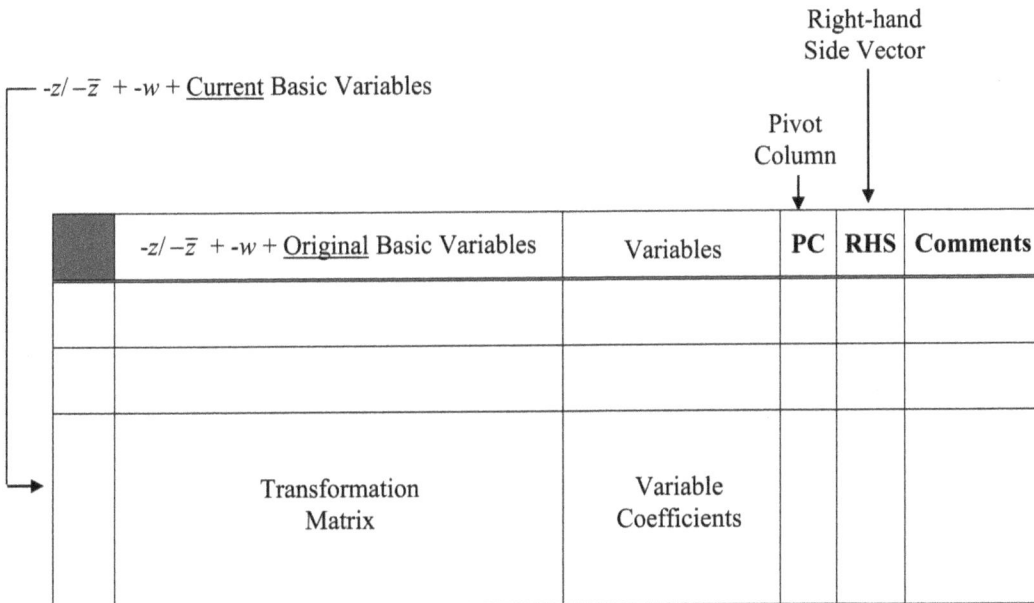

Right-hand
Side Vector

$-z / -\bar{z}\ +\ -w +$ Current Basic Variables

Pivot
Column

	$-z / -\bar{z}\ +\ -w +$ Original Basic Variables	Variables	PC	RHS	Comments
	Transformation Matrix	Variable Coefficients			

When using the tableau version of the two-phase revised simplex method it is not necessary to calculate and use the J matrix. Jordan elimination can be applied to the appropriate elements within the tableau in the same way that it is in the simplex methods.

Example 4.3

Use the two-phase revised simplex method to solve the following linear programming problem :

$$\begin{aligned}
\text{Minimise} \quad & z = -5x_1 - 2x_2 - 2x_3 \\
\text{Subject to :} \quad & x_1 + x_2 + x_3 = 1 \\
& x_1 + 2x_2 - x_3 = 2 \\
& x_1, x_2, x_3 \geq 0
\end{aligned}$$

Solution

Step 1 : Write the problem in canonical form, partition it, calculate the initial T matrix, write down the initial tableau and initialise the label i

Transposing the objective function and adding artificial variables a_1 and a_2 :

Minimise $-z - 5x_1 - 2x_2 - 2x_3 = 0$

Subject to : $x_1 + x_2 + x_3 + a_1 = 1$ ---- (1)

$x_1 + 2x_2 - x_3 + a_2 = 2$ ---- (2)

$x_1, x_2, x_3, a_1, a_2 \geq 0$

Let $w = a_1 + a_2$ i.e. $-w + a_1 + a_2 = 0$ ---- (3)

$$(1) + (2) \Rightarrow 2x_1 + 3x_2 + a_1 + a_2 = 3 \quad ---- (4)$$

$$(3) - (4) \Rightarrow -w - 2x_1 - 3x_2 = -3$$

Adding the w-row the linear programming problem can be written in canonical form as :

Minimise $-z - 5x_1 - 2x_2 - 2x_3 = 0$

$-w - 2x_1 - 3x_2 = -3$

Subject to : $x_1 + x_2 + x_3 + a_1 = 1$

$x_1 + 2x_2 - x_3 + a_2 = 2$

$x_1, x_2, x_3, a_1, a_2 \geq 0$

Partitioning this problem :

$$\underline{x}_B = \begin{bmatrix} a_1 \\ a_2 \end{bmatrix} \quad , \quad \underline{x}_N = \begin{bmatrix} x_1 \\ x_2 \\ x_3 \end{bmatrix}$$

$$\underline{c}_B = \begin{bmatrix} 0 & 0 \end{bmatrix} \quad , \quad \underline{c}_N = \begin{bmatrix} -5 & -2 & -2 \end{bmatrix}$$

$$\underline{d}_B = \begin{bmatrix} 0 & 0 \end{bmatrix} \quad , \quad \underline{d}_N = \begin{bmatrix} -2 & -3 & 0 \end{bmatrix}$$

$$B = \begin{bmatrix} 1 & 0 \\ 0 & 1 \end{bmatrix} \quad , \quad N = \begin{bmatrix} 1 & 1 & 1 \\ 1 & 2 & -1 \end{bmatrix}$$

In the two-phase algorithm the initial T matrix is given by :

$$T_0 = \begin{bmatrix} 1 & 0 & -\underline{c}_B B^{-1} \\ 0 & 1 & -\underline{d}_B B^{-1} \\ 0 & 0 & B^{-1} \end{bmatrix}$$

Here :

$$B^{-1} = \begin{bmatrix} 1 & 0 \\ 0 & 1 \end{bmatrix} \quad , \quad -\underline{c}_B B^{-1} = -\begin{bmatrix} 0 & 0 \end{bmatrix} \begin{bmatrix} 1 & 0 \\ 0 & 1 \end{bmatrix} = \begin{bmatrix} 0 & 0 \end{bmatrix}$$

$$-\underline{d}_B B^{-1} = -\begin{bmatrix} 0 & 0 \end{bmatrix} \begin{bmatrix} 1 & 0 \\ 0 & 1 \end{bmatrix} = \begin{bmatrix} 0 & 0 \end{bmatrix}$$

Hence :

$$T_0 = \begin{bmatrix} 1 & 0 & 0 & 0 \\ 0 & 1 & 0 & 0 \\ 0 & 0 & 1 & 0 \\ 0 & 0 & 0 & 1 \end{bmatrix}$$

The initial two-phase revised simplex tableau is :

	$-z$	$-w$	a_1	a_2	x_1	x_2	x_3	a_1	a_2	PC	RHS
$-z$	1	0	0	0	-5	-2	-2	0	0		0
$-w$	0	1	0	0	-2	-3	0	0	0		-3
a_1	0	0	1	0	1	1	1	1	0		1
a_2	0	0	0	1	1	2	-1	0	1		2

Let $i = 1$

Phase 1

Step 2 : Find the pivot variable :

$$d_1^{(i)} = d_1^{(1)} = \underline{T}_2 \underline{x}_1 = -2$$

Since $d_1^{(1)} < 0$, x_1 is the pivot variable $i.e.$ x_1 will join the basis

Second row of T matrix

$\underline{T}_2$

	$-z$	$-w$	a_1	a_2	x_1	x_2	x_3	a_1	a_2	PC	RHS
$-z$	1	0	0	0	-5	-2	-2	0	0		0
$-w$	0	1	0	0	-2	-3	0	0	0		-3
a_1	0	0	1	0	1	1	1	1	0		1
a_2	0	0	0	1	1	2	-1	0	1		2

$\underline{x}_1 \quad \underline{x}_2 \quad \underline{x}_3$

Step 3 : Calculate the pivot column :

$T\underline{x}_1$

	$-z$	$-w$	a_1	a_2	x_1	x_2	x_3	a_1	a_2	PC	RHS
$-z$	1	0	0	0	-5	-2	-2	0	0	-5	0
$-w$	0	1	0	0	-2	-3	0	0	0	-2	-3
a_1	0	0	1	0	1	1	1	1	0	1	1
a_2	0	0	0	1	1	2	-1	0	1	1	2

Step 4 : Find the pivot row :

	−z	−w	a_1	a_2	x_1	x_2	x_3	a_1	a_2	PC	RHS	Ratio
−z	1	0	0	0	-5	-2	-2	0	0	-5	0	N/A
−w	0	1	0	0	-2	-3	0	0	0	-2	-3	N/A
a_1	0	0	1	0	1	1	1	1	0	1	1	1/1=1
a_2	0	0	0	1	1	2	-1	0	1	1	2	2/1=2

a_1 will leave the basis

Pivot

Smallest Ratio

Step 5 : Update the current basis labels :

	−z	−w	a_1	a_2	x_1	x_2	x_3	a_1	a_2	PC	RHS
−z	1	0	0	0	-5	-2	-2	0	0	-5	0
−w	0	1	0	0	-2	-3	0	0	0	-2	-3
x_1	0	0	1	0	1	1	1	1	0	1	1
a_2	0	0	0	1	1	2	-1	0	1	1	2

Label updated to indicate that x_1 has
joined the basis and a_1 has left the basis

Step 6 : Perform Jordan elimination on the right-hand side vector and the current T matrix :

	−z	−w	a_1	a_2	x_1	x_2	x_3	a_1	a_2	PC	RHS	Row Operations Performed
−z	1	0	5	0	-5	-2	-2		0		5	Row 1 + 5 x Pivot Row
−w	0	1	2	0	-2	-3	0		0		-1	Row 2 + 2 x Pivot Row
x_1	0	0	1	0	1	1	1		0		1	Pivot Row
a_2	0	0	-1	1	1	2	-1		1		1	Row 4 - Pivot Row

T matrix updated

Pivot column left blank as before because it
is recalculated in **Step 3** of the next iteration

Right-hand side
vector updated

Artificial variable a_1 discarded
because it has left the basis *i.e.* has
become non-basic

Step 7 : Let $i = i+1$ *i.e.* $i = 2$ and go back to Step 2.

The complete (and annotated) phase 1 revised simplex tableau for this problem is :

	$-z$	$-w$	a_1	a_2	x_1	x_2	x_3	a_1	a_2	PC	RHS	Comments
$-z$	1	0	0	0	-5	-2	-2	0	0	-5	0	$d_1^{(1)}<0$ *i.e.* x_1 joins basis
$-w$	0	1	0	0	-2	-3	0	0	0	-2	-3	
a_1	0	0	1	0	1	1	1	1	0	1	1	Smallest ratio =1 *i.e.* a_1 leaves basis
a_2	0	0	0	1	1	2	-1	0	1	1	2	a_1 can be discarded
$-z$	1	0	5	0	-5	-2	-2		0	3	5	$d_2^{(2)}<0$ *i.e.* x_2 joins basis
$-w$	0	1	2	0	-2	-3	0		0	-1	-1	
x_1	0	0	1	0	1	1	1		0	1	1	Smallest ratio =1 *i.e.* x_1 leaves basis
a_2	0	0	-1	1	1	2	-1		1	1	1	
$-z$	1	0	2	0	-5	-2	-2		0		2	$d_1^{(3)}>0, d_3^{(3)}>0$ *i.e.* phase 1 complete
$-w$	0	1	3	0	-2	-3	0		0		0	$w_{min}=0$ *i.e.* problem feasible
x_2	0	0	1	0	1	1	1		0		1	Begin phase 2
a_2	0	0	-2	1	1	2	-1		1		0	

The final tableau shows that $w_{min} = 0$ *i.e.* that the linear programming problem is feasible and that phase 2 can begin.

Phase 2
The initial phase 2 revised simplex tableau is :

	$-z$	$-w$	a_1	a_2	x_1	x_2	x_3	a_2	PC	RHS
$-z$	1	0	2	0	-5	-2	-2	0		2
$-w$	0	1	3	0	-2	-3	0	0		0
x_2	0	0	1	0	1	1	1	0		1
a_2	0	0	-2	1	1	2	-1	1		0

Step 2 : Find the pivot variable :

$$c_1^{(i)} = c_1^{(4)} = \underline{T}_1 \underline{x}_1 = -3$$

First row of T matrix   Since $c_1^{(4)} < 0$, x_1 is the pivot variable *i.e.* x_1 will join the basis

$\underline{T}_1$

	$-z$	$-w$	a_1	a_2	x_1	x_2	x_3	a_2	**PC**	**RHS**
$-z$	1	0	2	0	-5	-2	-2	0		2
$-w$	0	1	3	0	-2	-3	0	0		0
x_2	0	0	1	0	1	1	1	0		1
a_2	0	0	-2	1	1	2	-1	1		0

$\underline{x}_1$ $\underline{x}_2$ $\underline{x}_3$

Step 3 : Calculate the pivot column :

$T\underline{x}_1$

	$-z$	$-w$	a_1	a_2	x_1	x_2	x_3	a_2	**PC**	**RHS**
$-z$	1	0	2	0	-5	-2	-2	0	-3	2
$-w$	0	1	3	0	-2	-3	0	0	1	0
x_2	0	0	1	0	1	1	1	0	1	1
a_2	0	0	-2	1	1	2	-1	1	-1	0

Step 4 : Find the pivot row :

	$-z$	$-w$	a_1	a_2	x_1	x_2	x_3	a_2	**PC**	**RHS**	**Ratio**
$-z$	1	0	2	0	-5	-2	-2	0	-3	2	N/A
$-w$	0	1	3	0	-2	-3	0	0	1	0	0/1=0
x_2	0	0	1	0	1	1	1	0	1	1	1/1=1
a_2	0	0	-2	1	1	2	-1	1	-1	0	N/A

$-w$ will leave the basis   Pivot   Smallest Ratio

Step 5 : Update the current basis labels :

	$-z$	$-w$	a_1	a_2	x_1	x_2	x_3	a_2	PC	RHS	Ratio
$-z$	1	0	2	0	-5	-2	-2	0	-3	2	N/A
x_1	0	1	3	0	-2	-3	0	0	1	0	0/1=0
x_2	0	0	1	0	1	1	1	0	1	1	1/1=1
a_2	0	0	-2	1	1	2	-1	1	-1	0	N/A

Label updated to indicate that x_1 has
joined the basis and $-w$ has left the basis

Step 6 : Perform Jordan elimination on the right-hand side vector and the current T matrix :

	$-z$	$-w$	a_1	a_2	x_1	x_2	x_3	a_2	PC	RHS	Row Operations Performed
$-z$	1	3	11	0	-5	-2	-2	0		2	Row 1 + 3 x Pivot Row
x_1	0	1	3	0	-2	-3	0	0		0	Pivot Row
x_2	0	-1	-2	0	1	1	1	0		1	Row 3 - Pivot Row
a_2	0	1	1	1	1	2	-1	1		0	Row 4 + Pivot Row

T matrix updated   Pivot column left blank as before   Right-hand side vector updated

Step 7 : Let $i = i + 1$ *i.e.* $i = 5$ and go back to Step 2.

The complete (and annotated) two-phase revised simplex tableau for this problem is :

	$-z$	$-w$	a_1	a_2	x_1	x_2	x_3	a_1	a_2	PC	RHS	Comments
$-z$	1	0	0	0	-5	-2	-2	0	0	-5	0	$d_1^{(1)}<0$ i.e. x_1 joins basis
$-w$	0	1	0	0	-2	-3	0	0	0	-2	-3	
a_1	0	0	1	0	1	1	1	1	0	1	1	Smallest ratio =1 i.e. a_1 leaves basis
a_2	0	0	0	1	1	2	-1	0	1	1	2	a_1 can be discarded
$-z$	1	0	5	0	-5	-2	-2		0	3	5	$d_2^{(2)}<0$ i.e. x_2 joins basis
$-w$	0	1	2	0	-2	-3	0		0	-1	-1	
x_1	0	0	1	0	1	1	1		0	1	1	Smallest ratio =1 i.e. x_1 leaves basis
a_2	0	0	-1	1	1	2	-1		1	1	1	
$-z$	1	0	2	0	-5	-2	-2		0		2	$d_1^{(3)}>0, d_3^{(3)}>0$ i.e. phase 1 complete
$-w$	0	1	3	0	-2	-3	0		0		0	$w_{min}=0$ i.e. problem feasible
x_2	0	0	1	0	1	1	1		0		1	Begin phase 2
a_2	0	0	-2	1	1	2	-1		1		0	
$-z$	1	0	2	0	-5	-2	-2		0	-3	2	$c_1^{(4)}<0$ i.e. x_1 joins basis
$-w$	0	1	3	0	-2	-3	0		0	1	0	Smallest ratio =0 i.e. $-w$ leaves basis
x_2	0	0	1	0	1	1	1		0	1	1	
a_2	0	0	-2	1	1	2	-1		1	-1	0	
$-z$	1	3	11	0	-5	-2	-2		0		2	$c_3^{(5)}>0$
x_1	0	1	3	0	-2	-3	0		0		0	a_1 discarded, a_2 ignored
x_2	0	-1	-2	0	1	1	1		0		1	Solution optimal
a_2	0	1	1	1	1	2	-1		1		0	

The final tableau shows that the solution is now optimal. Reading across the tableau from left to right the optimal solution is $-z=2$ i.e. $z=-2$, $x_1=0$, $x_2=1$ and $x_3=0$.

4.8 Problem Cases
When using the revised simplex method and the two-phase revised simplex method problem cases such as non-unique solutions, infeasible problems, unbounded problems and degeneracy can be identified and dealt with in the same way as they are when using the ordinary simplex and two-phase simplex methods.

4.9 Exercises 4b

1. Use the two-phase revised simplex method to solve the following linear programming problems :

    (i)   Maximise $z = x_1 + x_2$

            Subject to : $4x_1 + 5x_2 \leq 200$

                         $x_1 + 4x_2 \geq 40$

                         $2x_1 + 3x_2 = 90$

                         $x_1, x_2 \geq 0$

    (ii)   Maximise $z = 5x_1 + 2x_2 + 2x_3 - 7$

            Subject to : $x_1 + x_2 + x_3 = 1$

                         $x_1 + 2x_2 - x_3 = 2$

                         $x_1, x_2, x_3 \geq 0$

    (iii)   Maximise $z = 1/2 x_1 - x_2$

            Subject to : $x_1 + x_2 \geq 1$

                         $x_2 \leq 4$

                         $x_1 + 2x_2 \leq 6$

                         $x_1 - 2x_2 \leq 2$

                         $x_1, x_2 \geq 0$

2. The linear programming problem from question 1(iii) has a non-unique solution. How can this be seen from the optimal solution ?

3. Show that the linear programming problem :

$$\begin{bmatrix} 1 & 0 & \underline{c}_B & \underline{c}_N \\ 0 & 1 & \underline{d}_B & \underline{d}_N \\ 0 & 0 & B & N \end{bmatrix} \begin{bmatrix} -z \\ -w \\ x_B \\ \underline{x}_N \end{bmatrix} = \begin{bmatrix} -c_0 \\ -w_0 \\ \underline{b} \end{bmatrix}$$

can be solved for the basic variables by premultiplying each side of this expression by a transformation matrix in the form :

$$T = \begin{bmatrix} 1 & 0 & -\underline{c}_B B^{-1} \\ 0 & 1 & -\underline{d}_B B^{-1} \\ \underline{0} & \underline{0} & B^{-1} \end{bmatrix}$$

Hint : Use the procedure from question 3 in Exercises 4a.

4. The manager of an investment fund buys bonds, preference shares and ordinary shares in order to make a profit for her investors. She is given the following table that shows the yields and risk factors involved :

	Yield (%)	Risk Factor
Bonds	6	1
Preference Shares	7	2
Ordinary Shares	12	4

(i) Formulate a linear programming model from which the manager can determine how to invest the fund if the risk factor must be at most 3 and the yield must be at least 7%.

(ii) Solve your linear programming model from 4(i) using the two-phase revised simplex method and interpret the solution you obtain.

5. Practical Difficulties Encountered When Solving Large Linear Programming Problems

5.1 Introduction
The linear programming problems that arise in commerce and industry are usually too large to be solved by hand and must therefore be solved by computer. Computer programs for solving linear programming problems must deal with the issues of **sparsity**, **rounding errors** and **termination**.

5.2 Sparsity
In Chapter 4 it was shown that linear programming problems can be written in matrix form. These matrices are usually sparse. This means that most (*i.e.* 95% or more) of the elements in the matrix are zero. When solving large linear programming problems it is wasteful of computer memory to store the zero elements. To overcome this problem computer programs use special sparse matrix storage schemes. A simple and commonly used sparse matrix storage scheme is Coordinate List. In this system a sparse matrix is stored as three vectors, one to hold the non-zero values, one to hold the row positions of the non-zero values and one to hold the column positions of the non-zero values and the number of non-zero values. The entries are usually sorted by row index and then column index. For example, consider the following sparse matrix :

$$\begin{bmatrix} 0 & 0 & 0 & 0 & 0 & 0 & 0 & 0 & 0 & 1 \\ 0 & 3 & 0 & 0 & 0 & 0 & 0 & 0 & 0 & 0 \\ 0 & 0 & 0 & 0 & 0 & 0 & 0 & 0 & 0 & 0 \\ 0 & 0 & -2 & 0 & 0 & 0 & 0 & 0 & 0 & 0 \\ 0 & 0 & 0 & 0 & 0 & 0 & 0 & 0 & 0 & 0 \\ 0 & 0 & 0 & 0 & 0 & 0 & 0 & 0 & 0 & 0 \\ 0 & 0 & 0 & 0 & 0 & 0 & 0 & 0 & 0 & 0 \\ 0 & 0 & 0 & 0 & 0 & 0 & 0 & 0 & 0 & 0 \\ 0 & 0 & 0 & 0 & 0 & 0 & 0 & 0 & 0 & 0 \\ 0 & 0 & 0 & 0 & 0 & 0 & 0 & 0 & 8 & 5 \end{bmatrix}$$

Using Coordinate List this matrix would be stored as three vectors as follows :

$$\text{Values} = \begin{bmatrix} 1 & 3 & -2 & 8 & 5 \end{bmatrix}$$

$$\text{Row Index} = \begin{bmatrix} 1 & 2 & 4 & 10 & 10 \end{bmatrix} \quad \text{Number of non-zero values}$$

$$\text{Column Index} = \begin{bmatrix} 10 & 2 & 3 & 9 & 10 & 5 \end{bmatrix}$$

5.3 Rounding Errors
Computers can store numbers to a fixed degree of accuracy only. Hence, each calculation introduces a rounding error into the solution. When solving large linear programming problems thousands of calculations are performed and the growth of these errors can seriously affect the accuracy of the solutions obtained. The methods most commonly used for minimising the effects of rounding errors in linear programming are scaling and reinversion.

Scaling

The growth of rounding errors can be reduced by scaling the problem before the solution is found so that all of the coefficients have similar magnitude. This operation is performed as follows :

- If the coefficients in the objective function or a constraint are significantly larger/smaller than those in the rest of the problem, divide/multiply through that row by an appropriate constant. Since this operation is applied to both sides of the problem it does not affect the solution of the linear programming problem.

- If the coefficients of a particular variable *e.g.* x_j are significantly larger/smaller than those of the other variables, replace x_j with a new variable kx'_j where k is the appropriate scale factor. Since this operation is not applied to both sides of the problem it does affect the solution of the linear programming problem. Hence, once the optimal solution of the problem has been found the optimal value of x_j must be calculated using the optimal value of x'_j.

Example 5.1

Consider the following linear programming problem :

$$\begin{aligned}
\text{Minimise} \quad & z = 0.2x_1 + 85x_2 - 0.1x_3 - 0.7x_4 \\
\text{Subject to :} \quad & 0.5x_1 + 200x_2 + 0.7x_3 - 0.8x_4 = 100 \\
& -0.3x_1 + 75x_2 - 1.0x_3 + 0.6x_4 \geq 30 \\
& 0.0006x_1 - 0.05x_2 - 0.0005x_3 - 0.001x_4 \geq 0.05 \\
& 0.1x_1 - 150x_2 + 0.3x_3 + 0.2x_4 \leq 85 \\
& x_1, x_2, x_3, x_4 \geq 0
\end{aligned}$$

Firstly, it can be seen that the coefficients in the third constraint are significantly smaller than those in the objective function and the other constraints. To overcome this problem, multiply through that constraint by 1000. The linear programming problem then becomes :

$$\begin{aligned}
\text{Minimise} \quad & z = 0.2x_1 + 85x_2 - 0.1x_3 - 0.7x_4 \\
\text{Subject to :} \quad & 0.5x_1 + 200x_2 + 0.7x_3 - 0.8x_4 = 100 \\
& -0.3x_1 + 75x_2 - 1.0x_3 + 0.6x_4 \geq 30 \\
& 0.6x_1 - 50x_2 - 0.5x_3 - 1.0x_4 \geq 50 \\
& 0.1x_1 - 150x_2 + 0.3x_3 + 0.2x_4 \leq 85 \\
& x_1, x_2, x_3, x_4 \geq 0
\end{aligned}$$

Finally, it can be seen that the coefficients of x_2 are significantly larger than those of the other variables. To overcome this problem, replace x_2 with a new variable $x'_2/200$. This divides all of the coefficients in the x_2 column by 200 and produces the following linear programming problem :

$$\text{Minimise} \quad z = 0.2x_1 + 85(x_2'/200) - 0.1x_3 - 0.7x_4$$
$$\text{Subject to:} \quad 0.5x_1 + 200(x_2'/200) + 0.7x_3 - 0.8x_4 = 100$$
$$-0.3x_1 + 75(x_2'/200) - 1.0x_3 + 0.6x_4 \geq 30$$
$$0.6x_1 - 50(x_2'/200) - 0.5x_3 - 1.0x_4 \geq 50$$
$$0.1x_1 - 150(x_2'/200) + 0.3x_3 + 0.2x_4 \leq 85$$
$$x_1, x_2', x_3, x_4 \geq 0$$

i.e.

$$\text{Minimise} \quad z = 0.2x_1 + 0.425x_2' - 0.1x_3 - 0.7x_4$$
$$\text{Subject to:} \quad 0.5x_1 + x_2' + 0.7x_3 - 0.8x_4 = 100$$
$$-0.3x_1 + 0.375x_2' - 1.0x_3 + 0.6x_4 \geq 30$$
$$0.6x_1 - 0.25x_2' - 0.5x_3 - 1.0x_4 \geq 50$$
$$0.1x_1 - 0.75x_2' + 0.3x_3 + 0.2x_4 \leq 85$$
$$x_1, x_2', x_3, x_4 \geq 0$$

The optimal value of x_2 can be then obtained from the optimal value of x_2' using the relationship :

$$x_2 = x_2'/200$$

Note : In this linear programming problem all of the coefficients have absolute values in the range [0.1, 1.0]. Hence this problem is less likely to be effected by rounding errors than the original one.

Reinversion
Reinversion is a rounding error reduction technique that can be incorporated into the simplex and revised simplex methods. The technique is usually applied every t iterations, $t \gg m$, where m is the number of structural constraints in the linear programming problem.

In the simplex methods reinversion involves identifying the current basis, returning to the initial tableau and reintroducing the appropriate variables into the basis, one at a time, in the appropriate order, overriding the usual pivot column selection rule. This process produces a more accurate (*i.e.* less rounded) tableau for the next iteration.

In the revised simplex methods reinversion involves identifying the current basis, repartitioning the original problem accordingly and then calculating a new T matrix using the appropriate formula.

If the linear programming problem contains type 1 (*i.e.* $\leq$) constraints only :

$$T = \begin{bmatrix} 1 & -\underline{c}_B B^{-1} \\ \underline{0} & B^{-1} \end{bmatrix}$$

If the linear programming problem contains a variety of constraint types :

$$T = \begin{bmatrix} 1 & 0 & -\underline{c}_B B^{-1} \\ 0 & 1 & -\underline{d}_B B^{-1} \\ \underline{0} & \underline{0} & B^{-1} \end{bmatrix} \text{ during phase 1 and } T = \begin{bmatrix} 1 & -\underline{c}_B B^{-1} \\ \underline{0} & B^{-1} \end{bmatrix} \text{ during phase 2.}$$

This process produces a more accurate (*i.e.* less rounded) T matrix for the next iteration.

Note : Reinversion involves additional calculations and is used only when solving large linear programming problems that are particularly sensitive to the effects of rounding errors.

Example 5.2
Consider the following linear programming problem :

$$\text{Minimise} \quad z = -10x_1 + 57x_2 + 9x_3 + 24x_4$$
$$\text{Subject to :} \quad x_1 - 11x_2 - 5x_3 + 18x_4 \leq 0$$
$$x_1 - 3x_2 - x_3 + 2x_4 \leq 0$$
$$x_1 \leq 1$$
$$x_1, x_2, x_3, x_4 \geq 0$$

Introducing slack variables x_5, x_6 and x_7 this problem can be written in canonical form as :

$$\text{Minimise} \quad -z - 10x_1 + 57x_2 + 9x_3 + 24x_4 = 0$$
$$\text{Subject to :} \quad x_1 - 11x_2 - 5x_3 + 18x_4 + x_5 = 0$$
$$x_1 - 3x_2 - x_3 + 2x_4 + x_6 = 0$$
$$x_1 + x_7 = 1$$
$$x_1, x_2, x_3, x_4, x_5, x_6, x_7 \geq 0$$

The Simplex Method
The initial simplex tableau is :

$-z$	x_1	x_2	x_3	x_4	x_5	x_6	x_7	RHS
1.0000	-10.0000	57.0000	9.0000	24.0000	0.0000	0.0000	0.0000	0.0000
0.0000	1.0000	-11.0000	-5.0000	18.0000	1.0000	0.0000	0.0000	0.0000
0.0000	1.0000	-3.0000	-1.0000	2.0000	0.0000	1.0000	0.0000	0.0000
0.0000	1.0000	0.0000	0.0000	0.0000	0.0000	0.0000	1.0000	1.0000

Suppose that after $t \gg 3$ iterations the following tableau is obtained :

$-z$	x_1	x_2	x_3	x_4	x_5	x_6	x_7	RHS
1.0000	0.0000	26.9989	-1.0020	44.0001	0.0000	10.0008	0.0000	0.0009
0.0000	1.0000	-3.0012	-1.0007	1.9996	0.0000	1.0005	0.0000	0.0012
0.0000	0.0000	-8.0100	-3.9994	15.9994	1.0000	-1.0008	0.0000	0.0011
0.0000	0.0000	2.9999	1.0003	-2.0002	0.0000	-1.0010	1.0000	1.0001

Reading this tableau from top to bottom it can be seen that the current basis is x_1, x_5 and x_7.

Returning to the initial tableau and reintroducing these variables into the basis, one at a time, in that order, overriding the usual pivot column selection rule produces the following tableau :

$-z$	x_1	x_2	x_3	x_4	x_5	x_6	x_7	RHS
1.0000	0.0000	27.0000	-1.0000	44.0000	0.0000	10.0000	0.0000	0.0000
0.0000	1.0000	-3.0000	-1.0000	2.0000	0.0000	1.0000	0.0000	0.0000
0.0000	0.0000	-8.0000	-4.0000	16.0000	1.0000	-1.0000	0.0000	0.0000
0.0000	0.0000	3.0000	1.0000	-2.0000	0.0000	-1.0000	1.0000	1.0000

This tableau is more accurate (*i.e.* less rounded) than the one above and is used in the next iteration.

The Revised Simplex Method

Suppose again that after $t \gg 3$ iterations the current basis is x_1, x_5 and x_7. Since the linear programming problem contains type 1 constraints only :

$$T = \begin{bmatrix} 1 & -\underline{c}_B B^{-1} \\ \underline{0} & B^{-1} \end{bmatrix}$$

Repartitioning the original problem :

$$\underline{x}_B = \begin{bmatrix} x_1 \\ x_5 \\ x_7 \end{bmatrix}, \quad \underline{c}_B = \begin{bmatrix} -10 & 0 & 0 \end{bmatrix}, \quad B = \begin{bmatrix} 1 & 1 & 0 \\ 1 & 0 & 0 \\ 1 & 0 & 1 \end{bmatrix}$$

Hence :
$$B^{-1} = \begin{bmatrix} 0 & 1 & 0 \\ 1 & -1 & 0 \\ 0 & -1 & 1 \end{bmatrix}, \quad -\underline{c}_B B^{-1} = -\begin{bmatrix} -10 & 0 & 0 \end{bmatrix} \begin{bmatrix} 0 & 1 & 0 \\ 1 & -1 & 0 \\ 0 & -1 & 1 \end{bmatrix} = \begin{bmatrix} 0 & 10 & 0 \end{bmatrix}$$

Substituting :

$$T = \begin{bmatrix} 1 & 0 & 10 & 0 \\ 0 & 0 & 1 & 0 \\ 0 & 1 & -1 & 0 \\ 0 & 0 & -1 & 1 \end{bmatrix}$$

This T matrix is more accurate (*i.e.* less rounded) than the one obtained by applying $t \gg 3$ iterations of the usual algorithm and is used in the next iteration.

Supplementary Exercise

Write the linear programming problem from Example 5.2 in matrix form. Premultiply each side of this expression by the T matrix above and show that the accurate simplex tableau is obtained.

5.4 Termination

In the two-phase simplex and revised simplex methods phase 1/phase 2 ends when all of the coefficients in the w-row/z-row are greater than or equal to zero. When these methods are implemented on a computer rounding errors can sometimes obscure the true values of these coefficients with serious consequences. For example, suppose that a particular coefficient is theoretically feasible/optimal with a small positive value but that due to the effects of rounding errors it is stored with a small negative value. If all of the other coefficients in the w-row/z-row are positive, this column will then be chosen as the pivot column causing the algorithm to move away from the feasible/optimal solution. To overcome this problem it is usual to terminate phase 1/phase 2 of these methods when the solution is nearly feasible/optimal *i.e.* when the values of the coefficients in the w-row/z-row are greater than a small negative value *e.g.* -10^{-4}.

5.5 Exercises 5

1. Consider the following sparse matrix :

$$\begin{bmatrix}
0 & 0 & 0 & 2 & 0 & 0 & 0 & 0 & 0 & 0 & 0 & 0 & 0 & 0 & 1 \\
4 & 0 & 0 & 0 & 0 & 0 & 0 & 0 & 0 & 0 & 5 & 0 & 0 & 0 & 0 \\
0 & 0 & 0 & 0 & 0 & 0 & 0 & 0 & 0 & 0 & 0 & 0 & 0 & 0 & 0 \\
0 & 0 & 0 & 0 & 0 & 0 & 1 & 0 & 0 & 0 & 0 & 0 & 0 & 0 & 0 \\
0 & 0 & 0 & 0 & 0 & 0 & 0 & 0 & 0 & 0 & 0 & 0 & 0 & 0 & 0 \\
0 & 0 & 0 & 0 & 0 & 0 & 0 & 0 & 0 & 0 & 0 & 2 & 0 & 0 & 0 \\
0 & 0 & 0 & 0 & 0 & 0 & 0 & 0 & 0 & 0 & 0 & 0 & 0 & 0 & 0 \\
0 & 1 & 0 & 0 & 0 & 0 & 0 & 0 & 0 & 0 & 0 & 0 & 0 & 0 & 0 \\
0 & 0 & 0 & 0 & 0 & 0 & 0 & 0 & 0 & 0 & 0 & 7 & 0 & 0 & 0 \\
0 & 0 & 0 & 0 & 0 & 0 & 0 & 0 & 0 & 0 & 0 & 0 & 0 & 0 & 0 \\
0 & 0 & 2 & 0 & 0 & 0 & 0 & 0 & 0 & 0 & 0 & 0 & 0 & 0 & 0 \\
0 & 0 & 0 & 0 & 0 & 0 & 0 & 0 & 0 & 0 & 0 & 0 & 0 & 0 & 0 \\
0 & 0 & 0 & 0 & 0 & 0 & 0 & 0 & 0 & 0 & 0 & 0 & 0 & 0 & 0 \\
0 & 0 & 0 & 0 & 0 & 0 & 0 & 0 & 1 & 0 & 3 & 0 & 0 & 0 & 0 \\
0 & 0 & 0 & 0 & 0 & 0 & 0 & 0 & 0 & 0 & 0 & 0 & 0 & 0 & 0
\end{bmatrix}$$

Show how this matrix would be stored using the Coordinate List sparse matrix storage scheme.

2. Consider the following linear programming problem :

Minimise $z = 4x_1 + 1346x_2 - 0.001x_3$

Subject to : $2x_1 + 5614x_2 + 0.005x_3 \le 2746.01$

$-x_1 + 6594x_2 + 0.004x_3 = 3037.40$

$0.007x_1 + 3x_2 + 0.000006x_3 \le 1.82$

$x_1, x_2, x_3 \ge 0$

(i) Scale this problem so that the coefficients have similar magnitude.

(ii) Solve the scaled problem from 2(i) using the two-phase simplex method. Work to 4 decimal places.

(iii) Use the optimal solution of the scaled problem to write down the optimal solution of the original problem.

3. Consider the following linear programming problem :

Minimise $\quad z = 0.93x_1 - 2.00x_2$

Subject to : $\quad 0.90x_1 + 0.85x_2 \geq 2.30$

$\quad\quad\quad\quad -1.10x_1 + 0.96x_2 \geq 1.20$

$\quad\quad\quad\quad 0.92x_2 \leq 3.00$

$\quad\quad\quad\quad x_1, x_2 \geq 0$

Suppose that after $t \gg 3$ iterations of the two-phase simplex method and working to 4 decimal places, the following (phase 2) tableau was obtained :

-z	x_1	x_2	x_3	x_4	x_5	RHS
1.0000	0.0000	0.0000	0.0000	-0.8400	1.3100	4.8800
0.0000	1.0000	0.0000	0.0000	0.9300	0.9400	1.7200
0.0000	0.0000	1.0000	0.0000	0.0100	1.0800	3.2500
0.0000	0.0000	0.0000	1.0000	0.8100	1.7700	2.0900

where x_3 and x_4 are surplus variables and x_5 is a slack variable. The artificial variables became non-basic during phase 1 and were discarded. The w-row became trivial and was also discarded.

(i) Write the linear programming problem above in canonical form then discard the w-row and the artificial variables. Write the reduced problem in tableau form.

(ii) Identify the current basis from the tableau above.

(iii) Apply reinversion to the tableau from 3(i). Work to 4 decimal places.

4. (i) Calculate the (new) current T matrix for the linear programming problem from question 3. Work to 4 decimal places.

(ii) Write the reduced linear programming problem from question 3(i) in matrix form.

(iii) Premultiply each side of this expression by the T matrix from 4(i) and show that the accurate simplex tableau from question 3(iii) is obtained. Work to 4 decimal places.

6. Other Variable and Constraint Types

6.1 Introduction
The simplex methods and the revised simplex methods require the structural variables to be non-negative (*i.e.* ≥ 0). However, linear programming problems arise that contain other variable types. These must be converted into non-negative variables before the solution can be found.

6.2 Non-Positive Variables
If a structural variable x_i is constrained to be **non-positive** (*i.e.* ≤ 0) then each occurrence of x_i must be replaced by a non-negative variable x_i' that is defined as :

$$x_i = -x_i'$$

6.3 Unconstrained Variables
If a structural variable x_i is **unconstrained** (*i.e.* can be positive, negative or zero) then each occurrence of x_i must be replaced by a pair of non-negative variables x_i' and x_i'' that are defined as:

$$x_i = x_i' - x_i''$$

Once the optimal solution of the modified linear programming problem has been found the optimal value of x_i can be found by substituting the optimal values of x_i' and x_i'' into the appropriate formula above.

Example
Consider the following linear programming problem :

$$
\begin{aligned}
\text{Minimise} \quad & z = 4x_1 - 6x_2 - x_3 \\
\text{Subject to :} \quad & 2x_1 + x_2 - x_3 \leq 6 \\
& 3x_1 - 2x_2 \geq 7 \\
& x_1 + x_2 + 2x_3 = 11 \\
& x_1 \geq 0, x_2 \leq 0, x_3 \text{ unconstrained}
\end{aligned}
$$

Rewrite this problem so that all of the structural variables are constrained to be non-negative.

Solution
The non-positive variable x_2 must be replaced by a non-negative variable x_2' that is defined as :

$$x_2 = -x_2'$$

The unconstrained variable x_3 must be replaced by a pair of non-negative variables x_3' and x_3'' that are defined as :

$$x_3 = x_3' - x_3''$$

Hence, the linear programming problem can be written as :

$$\begin{array}{ll} \text{Minimise} & z = 4x_1 - 6(-x_2') - (x_3' - x_3'') \\ \text{Subject to :} & 2x_1 + (-x_2') - (x_3' - x_3'') \le 6 \\ & 3x_1 - 2(-x_2') \ge 7 \\ & x_1 + (-x_2') + 2(x_3' - x_3'') = 11 \\ & x_1, x_2', x_3', x_3'' \ge 0 \end{array}$$

i.e.

$$\begin{array}{ll} \text{Minimise} & z = 4x_1 + 6x_2' - x_3' + x_3'' \\ \text{Subject to :} & 2x_1 - x_2' - x_3' + x_3'' \le 6 \\ & 3x_1 + 2x_2' \ge 7 \\ & x_1 - x_2' + 2x_3' - 2x_3'' = 11 \\ & x_1, x_2', x_3', x_3'' \ge 0 \end{array}$$

Note
The optimal solution of the modified problem is :

$$x_1 = 7/3, \ x_2' = 0, \ x_3' = 13/3, \ x_3'' = 0 \ \text{ and } \ z = 5$$

Hence, the optimal solution of the original problem is :

$$x_1 = 7/3, \ x_2 = 0, \ x_3 = 13/3 - 0 \ \textit{i.e.} \ x_3 = 13/3 \ \text{ and } \ z = 4\text{x}(7/3) - 6\text{x}(0) - (13/3) \ \textit{i.e.} \ z = 5$$

6.4 Binary Variables
A **binary variable** is one that can take the values **zero** and **one** only. Variables of this type are used for modelling **mutually exclusive** and **contingent** events. Two events A and B are said to be mutually exclusive if when A occurs, B cannot and vice-versa. Two events C and D are said to be contingent if when D can occur if and only if C does and vice-versa.

Notes
- To indicate that x_i is a binary variable, the following constraints and condition must be added to the linear programming problem :

$$\begin{array}{l} x_i \le 1 \\ x_i \ge 0 \\ x_i \ \text{ integer} \end{array}$$

- To indicate that <u>exactly **s**</u> of the **t** binary variables $x_1, x_2, \ldots, x_t$ must take the value one, the following constraint must be added to the linear programming problem :

$$\sum_{i=1}^{t} x_i = s$$

- To indicate that <u>at most</u> s of the t binary variables $x_1, x_2, \ldots, x_t$ can take the value one, the following constraint must be added to the linear programming problem :

$$\sum_{i=1}^{t} x_i \leq s$$

- To indicate that <u>at least</u> s of the t binary variables $x_1, x_2, \ldots, x_t$ must take the value one, the following constraint must be added to the linear programming problem :

$$\sum_{i=1}^{t} x_i \geq s$$

Example 6.1

Frederick Johnson Engineering is considering expanding its production and storage facilities. It has identified two towns as being suitable, Amersham and Barnsley. Its capital budget will allow it to build either a factory in Amersham or a factory in Barnsley or a factory in both towns. If the company build one factory only then it has enough capital to build a warehouse. However, for logistical reasons the warehouse will have to be built in the town where it has built the factory. The company has allocated £10,000,000 for this expansion. The table below gives the capital requirement for each factory and warehouse together with the estimated value of each property upon completion :

	Capital Requirement (£m)	Value Upon Completion (£m)
Factory in Amersham	6	9
Factory in Barnsley	3	5
Warehouse in Amersham	5	6
Warehouse in Barnsley	2	4

Formulate a linear programming model from which the management at Frederick Johnson can determine the building strategy that will maximise the total value of the properties upon completion.

Solution

In this problem the structural variables must model the decisions to either build or not build each factory and warehouse. Let :

z be the total value of the properties upon completion.

x_1 model the decision to build a factory at Amersham.

x_2 model the decision to build a factory at Barnsley.

x_3 model the decision to build a warehouse at Amersham.

x_4 model the decision to build a warehouse at Barnsley.

Here x_1, x_2, x_3 and x_4 are binary variables. A value of zero will mean "do not build" and a value of one will mean "build".

From the information given :

The total value of the properties upon completion is :

$$z = 9x_1 + 5x_2 + 6x_3 + 4x_4$$

The capital constraint is :

$$6x_1 + 3x_2 + 5x_3 + 2x_4 \leq 10$$

Since at most one warehouse can be built (*i.e.* since <u>at most</u> one of the two binary variables x_3, x_4 can take the value one) :

$$\sum_{i=3}^{4} x_i \leq 1$$

i.e. $x_3 + x_4 \leq 1$

Since a warehouse can be built at Amersham only if a factory is built there :

$$x_3 \leq x_1$$

i.e. $-x_1 + x_3 \leq 0$

Similarly, since a warehouse can be built at Barnsley only if a factory is built there :

$$x_4 \leq x_2$$

i.e. $-x_2 + x_4 \leq 0$

Since x_1, x_2, x_3 and x_4 are binary variables :

$$x_1, x_2, x_3, x_4 \leq 1$$
$$x_1, x_2, x_3, x_4 \geq 0$$
$$x_1, x_2, x_3, x_4 \text{ integer}$$

Combining these expressions, the required linear programming model becomes :

Maximise $z = 9x_1 + 5x_2 + 6x_3 + 4x_4$
Subject to : $6x_1 + 3x_2 + 5x_3 + 2x_4 \leq 10$
$x_3 + x_4 \leq 1$
$-x_1 + x_3 \leq 0$
$-x_2 + x_4 \leq 0$
$x_1, x_2, x_3, x_4 \leq 1$
$x_1, x_2, x_3, x_4 \geq 0$
$x_1, x_2, x_3, x_4 \text{ integer}$

Note

The procedures used for solving linear programming problems containing integer constrained variables are described in Chapter 8. However, using the Solver tool in *Excel*, the optimal solution of this linear programming problem is :

$$x_1 = 1, \ x_2 = 1, \ x_3 = 0, \ x_4 = 0 \text{ and } z = 14$$

Hence, Frederick Johnson Engineering should build a factory in Amersham and a factory in Barnsley. The total value of these properties upon completion will be £14m.

6.5 At Most Constraints

"At most" constraints are used to model situations in which "at most" n from a set of m, $(m > n)$ constraints can be true *i.e.* affect the optimal solution. For example, suppose that an engineering company has designs for five new products but has resources to manufacture three only. For illustration purposes, suppose that the linear programming model from which the company can determine its profit is :

Maximize	$z = 10x_1 + 20x_2$	- Total profit
Subject to :	$7x_1 + 4x_2 \leq 10$	- Product 1 constraint
	$11x_1 - 3x_2 \geq 6$	- Product 2 constraint
	$-12x_1 + 8x_2 \leq 14$	- Product 3 constraint
	$9x_1 + x_2 = 12$	- Product 4 constraint
	$x_1 + 6x_2 \leq 10$	- Product 5 constraint
	$x_1, x_2 \geq 0$	

The company will wish to know which three of these five products to manufacture in order to maximise its profit *i.e.* which three of these five constraints must be true. To ensure that the linear programming model will answer this question, the constraints above must be rewritten as "at most" constraints.

To rewrite a set of constraints as "at most" constraints :

1) An expression involving a binary variable must be added to the right-hand side of each constraint that will ensure that the constraint is <u>non-binding</u> (*i.e.* not affecting the optimal solution) when the binary variable has a value of <u>zero</u> and <u>binding</u> (*i.e.* affecting the optimal solution) when the binary variable has a value of <u>one</u>. The expression required depends upon the type of the constraint *i.e.*

- Each **type 1** constraint *i.e.* $\sum_{j=1}^{n} a_{ij}x_j \leq b_i$ must be rewritten as : $\sum_{j=1}^{n} a_{ij}x_j \leq b_i y_i + M(1 - y_i)$

- Each **type 3** constraint *i.e.* $\sum_{j=1}^{n} a_{ij}x_j \geq b_i$ must be rewritten as : $\sum_{j=1}^{n} a_{ij}x_j \geq b_i y_i$

- Each **type 2** constraint *i.e.* $\sum_{j=1}^{n} a_{ij}x_j = b_i$ must be replaced with a type 1 constraint and a type 3 constraint. Each of these constraints must then be rewritten as above. However, the **same** binary variable must be used in each constraint.

In these expressions y_i is the binary variable used in that constraint and M is a positive number (called the **penalty**) that is (significantly) larger than any other number in the linear programming problem. For example, in the manufacturing example a penalty value of 1000 would be appropriate.

2) The additional constraint $\sum_{i=1}^{n} y_i \leq n$ must be added to the linear programming problem. This constraint enforces the restriction that "at most" n constraints can be true.

For example, replacing the type 2 constraint $9x_1 + x_2 = 12$ with the constraints $9x_1 + x_2 \leq 12$, $9x_1 + x_2 \geq 12$ and using a penalty value of 1000, the linear programming model in the manufacturing example would be rewritten as :

$$\begin{aligned}
\text{Maximize} \quad & z = 10x_1 + 20x_2 \\
\text{Subject to :} \quad & 7x_1 + 4x_2 \leq 10y_1 + 1000(1 - y_1) \\
& 11x_1 - 3x_2 \geq 6y_2 \\
& -12x_1 + 8x_2 \leq 14y_3 + 1000(1 - y_3) \\
& 9x_1 + x_2 \leq 12y_4 + 1000(1 - y_4) \\
& 9x_1 + x_2 \geq 12y_4 \\
& x_1 + 6x_2 \leq 10y_5 + 1000(1 - y_5) \\
& y_1, y_2, y_3, y_4, y_5 \text{ binary, } x_1, x_2 \geq 0
\end{aligned}$$

This linear programming model would then be solved in the usual way. The optimal values of the binary variables would then provide the company with the information in requires. For example, if the optimal values were $y_1 = 0$, $y_2 = 1$, $y_3 = 0$, $y_4 = 1$ and $y_5 = 1$ then the company would conclude that its maximum profit would be obtained by manufacturing products 2, 4 and 5 *etc.*

6.6 Similar Constraint Types

- "At least" constraints are used to model situations in which "at least" n from a set of m, $(m > n)$ constraints must be true. These are formulated in exactly the same was as "at most" constraints except that the additional constraint added to the problem at step 2) is :

$$\sum_{i=1}^{n} y_i \geq n$$

- "Either/or" constraints are essentially "at most" constraints in which "at most" one from a set of two constraints can be true. For example, consider the following pair of constraints :

$$3x_1 + 11x_2 \leq 33 \quad \text{---- (1)}$$
$$x_1 + x_2 \leq 17 \quad \text{---- (2)}$$

To ensure that "either" constraint (1) can be true "or" constraint (2) can be true these constraints can be rewritten as :

$$3x_1 + 11x_2 \le 33y_1 + M(1 - y_1)$$
$$x_1 + x_2 \le 17y_2 + M(1 - y_2)$$
$$y_1 + y_2 \le 1$$

where y_1 and y_2 are binary variables and M is a penalty value as before. However, "either/or" constraints can be rewritten using a single binary variable. For example, the constraints above can be rewritten as :

$$3x_1 + 11x_2 \le 33 + My_1 \quad \text{---- (3)}$$
$$x_1 + x_2 \le 17 + M(1 - y_1) \quad \text{---- (4)}$$

The advantage of this formulation is that the linear programming problem containing these constraints is easier to solve than the linear programming problem containing the ones above since it contains one less variable and one less constraint.

Supplementary Exercises

1. Substitute $y_1 = 0$ into constraints (3) and (4) above. Which of these constraints is now binding and why ?

2. Substitute $y_1 = 1$ into constraints (3) and (4) above. Which of these constraints is now binding and why ?

Answers

1. When $y_1 = 0$, the constraints become :

$$3x_1 + 11x_2 \le 33 \quad \text{---- (5)}$$
$$x_1 + x_2 \le 17 + M \quad \text{---- (6)}$$

Since M is very large constraint (6) is now true for all values of x_1 and x_2 i.e. constraint (6) is non-binding i.e. has no affect on the solution.

2. $y_1 = 1$, the constraints become :

$$3x_1 + 11x_2 \le 33 + M \quad \text{---- (7)}$$
$$x_1 + x_2 \le 17 \quad \text{---- (8)}$$

Since M is very large constraint (7) is now true for all values of x_1 and x_2 i.e. constraint (7) is non-binding i.e. has no affect on the solution.

6.7 Exercises 6

1. Use the two-phase simplex method to solve the following linear programming problem :

$$\text{Maximise} \quad z = 2x_1 - x_2$$
$$\text{Subject to :} \quad x_1 + x_2 \leq 0.5$$
$$x_1 + 2x_2 \geq 1$$
$$-x_1 + 5x_2 \leq 5$$
$$2x_1 + x_2 \geq 0.5$$
$$x_1 \leq 0, \ x_2 \geq 0$$

2. Use the two-phase revised simplex method to solve the following linear programming problem :

$$\text{Minimise} \quad z = 2x_1 + 3x_2$$
$$\text{Subject to :} \quad x_1 + x_2 \leq -0.5$$
$$-2x_1 + 3x_2 \leq 2$$
$$7x_1 - 4x_2 \leq -7$$
$$x_1 \text{ unconstrained, } x_2 \geq 0$$

3. A company called Plastofactors manufactures plastic components for household goods and wishes to expand its product range. The directors of the company have identified five new components they would consider manufacturing but are able to allocate a four week period only (*i.e.* 20 working days) to evaluate the manufacturing problems associated with the new products. The table below shows the estimated value to the company of introducing each new component and the time each would take to evaluate by a senior engineer :

Component	Estimated Value (£000's)	Time to Evaluate (Days)
1	7	3
2	17	8
3	11	5
4	9	4
5	21	10

Production constraints dictate that if component 1 is to be manufactured then component 2 must also be manufactured. The marketing department advise the directors that components 3 and 4 will occupy the same market slot and so one of these should be produced only. Formulate a linear programming model from which the management at Plastofactors can determine the set of new components to manufacture in order to maximise the estimated value to the company. You do <u>not</u> need to solve your model.

4. (i)   Use the Solver tool in *Excel* to solve the following linear programming problem :

$$\text{Maximise} \quad z = 2x_1 + 3x_2$$

Subject to :
$$
\left.
\begin{aligned}
x_1 + x_2 &= 20 \\
x_1 - x_2 &\leq 5 \\
2x_1 + 3x_2 &\geq 25
\end{aligned}
\right\} \quad \text{At most two}
$$
$$x_1, x_2 \geq 0$$

Use a penalty value of 1000 in the "at most" constraints and interpret the solution you obtain. An example of using the Solver tool is given in Chapter 10.

(ii) Introduce the additional constraint $y_2 = 1$ into the linear programming problem from 4(i) and then resolve it. Here, y_2 is the binary variable used in the type 1 constraint. Interpret the solution you obtain.

5. (i)   Use the Solver tool in *Excel* to solve the following linear programming problem :

$$\text{Maximise} \quad z = x_1 + x_2$$

Subject to :
$$
\left.
\begin{aligned}
4x_1 + 5x_2 &\leq 200 \\
x_1 + 4x_2 &\geq 40 \\
2x_1 + 3x_2 &= 90
\end{aligned}
\right\} \quad \text{At least two}
$$
$$x_1, x_2 \geq 0$$

Use a penalty value of 1000 in the "at least" constraints and interpret the solution you obtain.

(ii) Introduce the additional constraint $y_3 = 1$ into the linear programming problem from 5(i) and then resolve it. Here, y_3 is the binary variable used in the type 2 constraint. Interpret the solution you obtain.

6. (i)   Use the Solver tool in *Excel* to solve the following linear programming problem ·

$$\text{Maximise} \quad z = 2x_1 + 3x_2$$

Subject to :
$$2x_1 + x_2 \leq 8$$
$$
\left.
\begin{aligned}
x_1 + x_2 &\leq 6 \\
x_1 + 12x_2 &\leq 10
\end{aligned}
\right\} \quad \text{Either/or}
$$
$$x_1, x_2 \geq 0$$

Use a penalty value of 1000 in the "either/or" constraints and interpret the solution you obtain.

(ii) Introduce the additional constraint $y_1 = 1$ into the linear programming problem from 6(i) and then resolve it. Here, y_1 is the binary variable used in the "either/or" constraints. Interpret the solution you obtain.

7. Duality

7.1 Introduction

Consider the following manufacturing example :

The Cloth Manufacturer's/Wool Buyer's Model

A company manufactures two types of cloth, Standard and Deluxe using three different colour wools *i.e.* red, green and yellow. Table 1 below shows the amount of each colour wool used to make one metre of each type of cloth together with the availability of each colour wool and the selling price per metre of each type of cloth.

	Standard	Deluxe	Availability
Red Wool	200g/m	300g/m	80000g
Green Wool	400g/m	600g/m	90000g
Yellow Wool	500g/m	700g/m	140000g
Selling Price	£6/m	£8/m	

Table 1

The company can sell all the cloth it produces. The manager of the company wishes to calculate the amount of each type of cloth to manufacture in order to maximise the total income. This problem can be formulated as a linear programming model.

Let :

z　be the total income from manufacturing and selling the cloth.

x_1　be the number of metres of Standard cloth produced.

x_2　be the number of metres of Deluxe cloth produced.

Then, the total income is :

$$z = 6x_1 + 8x_2$$

The total income is constrained by the availability of the wools. In each case the demand must be less than or equal to the availability. Hence :

The red wool constraint is :

$$200x_1 + 300x_2 \leq 80000$$

$$i.e. \quad 2x_1 + 3x_2 \leq 800$$

Similarly :

The green wool constraint is :

$$400x_1 + 600x_2 \leq 90000$$

$$i.e. \quad 4x_1 + 6x_2 \leq 900$$

The yellow wool constraint is :

$$500x_1 + 700x_2 \leq 140000$$

$$i.e. \quad 5x_1 + 7x_2 \leq 1400$$

Since the amount of each type of cloth produced cannot be negative the total income is also constrained by the trivial inequalities :

$$x_1, x_2 \geq 0$$

Combining these expressions the **Cloth Manufacturer's Model** is :

Maximise $\quad z = 6x_1 + 8x_2$

Subject to : $\quad 2x_1 + 3x_2 \leq 800$

$\qquad\qquad 4x_1 + 6x_2 \leq 900$

$\qquad\qquad 5x_1 + 7x_2 \leq 1400$

$\qquad\qquad x_1, x_2 \geq 0$

Instead of manufacturing and selling cloth the company could generate income by selling its entire stock of wool. Let :

v 	be the total income from selling the wool.

y_1 	be the selling price of red wool in £/100g.

y_2 	be the selling price of green wool in £/100g.

y_3 	be the selling price yellow wool in £/100g.

Then, the total income is :

Number of 100g units available

$$v = 800y_1 + 900y_2 + 1400y_3$$

Selling price in £/100g

Naturally, the <u>wool buyers</u> will wish to pay the least amount for the wool *i.e.* to minimise the value of v.

The company will not sell the wool unless the income it receives from the sale is **at least** as much as it would earn by manufacturing and selling the cloth.

The income the company will receive by selling the wool that will make one metre of the Standard cloth is :

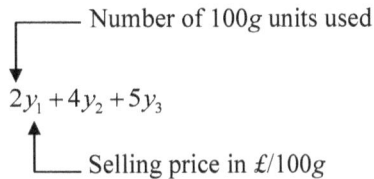

Number of 100g units used

$$2y_1 + 4y_2 + 5y_3$$

Selling price in £/100g

Since the Standard cloth sells for £6/m, the company will sell this wool only if :

$$2y_1 + 4y_2 + 5y_3 \geq 6$$

By the same argument the company will sell the wool that will make one metre of the Deluxe cloth only if :

$$3y_1 + 6y_2 + 7y_3 \geq 8$$

Since the selling price of each wool cannot be negative the total income is also constrained by the trivial inequalities :

$$y_1, y_2, y_3 \geq 0$$

Combining these expressions the **Wool Buyer's Model** is :

$$\text{Minimise} \quad v = 800y_1 + 900y_2 + 1400y_3$$
$$\text{Subject to :} \quad 2y_1 + 4y_2 + 5y_3 \geq 6$$
$$3y_1 + 6y_2 + 7y_3 \geq 8$$
$$y_1, y_2, y_3 \geq 0$$

By comparing this model with the Cloth Manufacturer's Model *i.e.*

$$\text{Maximise} \quad z = 6x_1 + 8x_2$$
$$\text{Subject to :} \quad 2x_1 + 3x_2 \leq 800$$
$$4x_1 + 6x_2 \leq 900$$
$$5x_1 + 7x_2 \leq 1400$$
$$x_1, x_2 \geq 0$$

a number of relationships can be seen :

- The constant values on the right-hand side of the Cloth Manufacturer's Model are the objective function coefficients in the Wool Buyer's Model (and vice-versa).

- The variable coefficients in the Cloth Manufacturer's Model are the constraint coefficients in the Wool Buyer's Model (and vice-versa).

- Since the income received by the company by selling the wool must be at least as much as it would earn by manufacturing and selling the cloth, the maximum value of z in the Cloth Manufacturer's Model is the minimum value of v in the Wool Buyer's Model.

These models illustrate the concept of **duality**. For every linear programming problem (in this context called the **primal**), there is a related linear programming problem (called the **dual**). Duality is an important concept in linear optimisation.

The **standard primal** problem is :

$$\text{Minimise} \quad z = \underline{c}\,\underline{x}$$
$$\text{Subject to :} \quad A\underline{x} \geq \underline{b}$$
$$\underline{x} \geq \underline{0}$$

where $\underline{c}$ is a row vector, $\underline{x}$ is a column vector, A is a matrix, and $\underline{b}$ is a column vector. Here, all of the constraints are **type 3** and the elements of the vector $\underline{b}$ can be positive, zero or negative. Notice that the objective function does not contain a constant c_0.

Using the relationships identified earlier the **dual** of the standard primal problem is :

$$\text{Maximise} \quad v = \underline{b}^T \underline{y}$$
$$\text{Subject to :} \quad A^T \underline{y} \leq \underline{c}^T$$
$$\underline{y} \geq \underline{0}$$

Note

Although the elements of the vector $\underline{b}$ can be negative in the standard primal problem, these elements must be greater than or equal to zero before the problem can be solved using the simplex methods or the revised simplex methods.

7.2 Converting a Linear Programming Problem to Standard Primal Form

The Objective Function

An objective function in the form $z = c_0 + \underline{c}\,\underline{x}$ can be converted to standard form by combining c_0 with one of the structural variables and introducing a new unconstrained variable. For example, suppose that the objective function is :

$$z = 3x_1 + 4x_2 - 1$$

This can be written as :

$$z = 3x_1 + 4\left(x_2 - 1/4\right)$$

Let $x_2' = x_2 - 1/4$ where x_2' is an unconstrained variable. Then, the objective function can be written as :

$$z = 3x_1 + 4x_2'$$

All occurrences of x_2 in the constraints must now be replaced with the expression $x_2' + 1/4$. For example, suppose that the linear programming problem contains the constraint :

$$5x_1 + 6x_2 \leq 3$$

Replacing x_2 with $x_2' + 1/4$ this can be written as :

$$5x_1 + 6\left(x_2' + 1/4\right) \le 3$$

$$5x_1 + 6x_2' + 3/2 \le 3$$

$$i.e. \quad 5x_1 + 6x_2' \le 3/2$$

Once the optimal solution of the linear programming problem has been found, the optimal value of x_2 can be found by substituting the optimal value of x_2' into the expression $x_2 = x_2' + 1/4$.

The Constraints
Each **type 1** constraint can be converted to standard form by multiplying through by -1. For example, consider the type 1 constraint :

$$4x_1 + 6x_2 - x_3 \le 5$$

Multiplying through by -1 this can be written as :

$$-4x_1 - 6x_2 + x_3 \ge -5$$

In the standard primal problem the term on the right-hand side can be positive, zero or negative.

Each **type 2** constraint can be converted to standard form by replacing it with a type 1 constraint <u>and</u> a type 3 constraint. The type 1 constraint can then be converted to a type 3 constraint using the procedure described above. For example, consider the type 2 constraint :

$$x_1 - 2x_2 + x_3 = 9$$

This can be replaced by the pair of constraints :

$$x_1 - 2x_2 + x_3 \le 9 \quad \text{---- (1)}$$
$$x_1 - 2x_2 + x_3 \ge 9 \quad \text{---- (2)}$$

Multiplying through constraint (1) by -1, the type 2 constraint can written as :

$$-x_1 + 2x_2 - x_3 \ge -9$$
$$x_1 - 2x_2 + x_3 \ge 9$$

7.3 Finding the Dual of a Linear Programming Problem
The dual of a linear programming problem can be found by converting the problem to standard primal form and then using the relationships identified earlier or by using the table method.

Using Standard Primal Form

Example 7.1
Convert the following linear programming problem to standard primal form and then write down the corresponding dual problem :

$$\text{Maximise} \quad z = 5x_1 - 3x_2 + 4x_3 - 2$$
$$\text{Subject to :} \quad x_1 + 2x_2 + 6x_3 \leq 50 \quad \text{---- (1)}$$
$$2x_1 - x_2 + x_3 = 16 \quad \text{---- (2)}$$
$$4x_1 - 2x_2 + 3x_3 \geq 20 \quad \text{---- (3)}$$
$$x_1 \geq 0, \ x_2 \leq 0, \ x_3 \text{ unconstrained}$$

Solution

The Objective Function
Converting to a minimisation problem the objective function can be written as :

$$\bar{z} = -5x_1 + 3x_2 - 4x_3 + 2$$

Removing the constant term :

$$\bar{z} = -5x_1 + 3x_2 - 4(x_3 - 1/2)$$

Let $x_3' = x_3 - 1/2$ where x_3' is an unconstrained variable. Then :

$$\bar{z} = -5x_1 + 3x_2 - 4x_3'$$

Reflecting this change into the constraints :

$$(1) \Rightarrow x_1 + 2x_2 + 6\left(x_3' + 1/2\right) \leq 50$$

$$i.e. \quad x_1 + 2x_2 + 6x_3' \leq 47$$

$$(2) \Rightarrow 2x_1 - x_2 + \left(x_3' + 1/2\right) = 16$$

$$i.e. \quad 2x_1 - x_2 + x_3' = 31/2$$

$$(3) \Rightarrow 4x_1 - 2x_2 + 3\left(x_3' + 1/2\right) \geq 20$$

$$i.e. \quad 4x_1 - 2x_2 + 3x_3' \geq 37/2$$

Combining these expressions the linear programming problem becomes :

$$\text{Minimise} \qquad \overline{z} = -5x_1 + 3x_2 - 4x_3'$$

$$\text{Subject to :} \quad x_1 + 2x_2 + 6x_3' \leq 47 \quad \text{---- } (1')$$

$$2x_1 - x_2 + x_3' = 31/2 \quad \text{---- } (2')$$

$$4x_1 - 2x_2 + 3x_3' \geq 37/2 \quad \text{---- } (3')$$

$$x_1 \geq 0, \ x_2 \leq 0, \ x_3' \text{ unconstrained}$$

The Constraints

Multiplying through constraint $(1')$ by -1 :

$$-x_1 - 2x_2 - 6x_3' \geq -47$$

Constraint $(2')$ can be replaced by the pair of constraints :

$$2x_1 - x_2 + x_3' \leq 31/2 \quad \text{---- } (4')$$

$$2x_1 - x_2 + x_3' \geq 31/2 \quad \text{---- } (5')$$

Multiplying through constraint $(4')$ by -1 :

$$-2x_1 + x_2 - x_3' \geq -31/2$$

Constraint $(3')$ is already a type 3 constraint.

Combining these expressions the linear programming problem can be written as :

$$\text{Minimise} \qquad \overline{z} = -5x_1 + 3x_2 - 4x_3'$$

$$\text{Subject to :} \quad -x_1 - 2x_2 - 6x_3' \geq -47$$

$$-2x_1 + x_2 - x_3' \geq -31/2$$

$$2x_1 - x_2 + x_3' \geq 31/2$$

$$4x_1 - 2x_2 + 3x_3' \geq 37/2$$

$$x_1 \geq 0, \ x_2 \leq 0, \ x_3' \text{ unconstrained}$$

The Variables

The non-positive variable x_2 must be replaced by a new non-negative variable x_2' where $x_2 = -x_2'$.

The unconstrained variable x_3' must be replaced by a pair of non-negative variables x_3'' and x_3''' where $x_3' = x_3'' - x_3'''$.

Hence, the standard primal form of the linear programming problem becomes :

Minimise
$$\bar{z} = -5x_1 - 3x_2' - 4\left(x_3'' - x_3'''\right)$$

Subject to :
$$-x_1 + 2x_2' - 6\left(x_3'' - x_3'''\right) \geq -47$$
$$-2x_1 - x_2' - \left(x_3'' - x_3'''\right) \geq -31/2$$
$$2x_1 + x_2' + \left(x_3'' - x_3'''\right) \geq 31/2$$
$$4x_1 + 2x_2' + 3\left(x_3'' - x_3'''\right) \geq 37/2$$
$$x_1, x_2', x_3'', x_3''' \geq 0$$

i.e. Minimise
$$\bar{z} = -5x_1 - 3x_2' - 4x_3'' + 4x_3'''$$

Subject to :
$$-x_1 + 2x_2' - 6x_3'' + 6x_3''' \geq -47$$
$$-2x_1 - x_2' - x_3'' + x_3''' \geq -31/2$$
$$2x_1 + x_2' + x_3'' - x_3''' \geq 31/2$$
$$4x_1 + 2x_2' + 3x_3'' - 3x_3''' \geq 37/2$$
$$x_1, x_2', x_3'', x_3''' \geq 0$$

Using the relationships identified earlier the dual of this primal problem is :

Maximise
$$v = -47y_1 - 31/2y_2 + 31/2y_3 + 37/2y_4$$

Subject to :
$$-y_1 - 2y_2 + 2y_3 + 4y_4 \leq -5 \quad \text{---- (1'')}$$
$$2y_1 - y_2 + y_3 + 2y_4 \leq -3 \quad \text{---- (2'')}$$
$$-6y_1 - y_2 + y_3 + 3y_4 \leq -4 \quad \text{---- (3'')}$$
$$6y_1 + y_2 - y_3 - 3y_4 \leq 4 \quad \text{---- (4'')}$$
$$y_1, y_2, y_3, y_4 \geq 0$$

Tidying up :

Multiplying through constraints $(1'')$, $(2'')$, and $(3'')$ by -1 to make the right-hand side terms positive :

$$(1'') \Rightarrow y_1 + 2y_2 - 2y_3 - 4y_4 \geq 5$$

$$(2'') \Rightarrow -2y_1 + y_2 - y_3 - 2y_4 \geq 3$$

$$(3'') \Rightarrow 6y_1 + y_2 - y_3 - 3y_4 \geq 4 \quad \text{---- (5'')}$$

Constraints $(4'')$ and $(5'')$ can be combined into the single type 2 constraint :

$$6y_1 + y_2 - y_3 - 3y_4 = 4$$

Hence, the dual linear programming problem becomes :

$$\text{Maximise} \quad v = -47y_1 - 31/2y_2 + 31/2y_3 + 37/2y_4$$

$$\text{Subject to :} \quad y_1 + 2y_2 - 2y_3 - 4y_4 \geq 5$$

$$-2y_1 + y_2 - y_3 - 2y_4 \geq 3$$

$$6y_1 + y_2 - y_3 - 3y_4 = 4$$

$$y_1, y_2, y_3, y_4 \geq 0$$

Using the Table Method

Once the constant term has been removed from the objective function the dual of a linear programming problem can be found using the Table 2 below. This table provides the quickest way of finding the dual problem as it does not require the linear programming problem to be converted to standard primal form.

Result	Primal		Dual
1	Minimise z	↔	Maximise v
2	$\underline{c}$	↔	$\underline{b}^T$
3	$\underline{b}$	↔	$\underline{c}^T$
4	A	↔	A^T
5	Type 1 *i.e.* ≤ constraint	↔	Non-positive variable
6	Type 2 *i.e.* = constraint	↔	Unconstrained variable
7	Type 3 *i.e.* ≥ constraint	↔	Non-negative variable
8	Non-negative variable	↔	Type 1 *i.e.* ≤ constraint
9	Non-positive variable	↔	Type 3 *i.e.* ≥ constraint
10	Unconstrained variable	↔	Type 2 *i.e.* = constraint

Table 2

The procedure for using this table is as follows :

- Use results **1** and **2** to write down the dual objective function.

- Use result **4** to write down the left-hand sides of the dual constraints.

- Use result **3** to write down the right-hand sides of the dual constraints.

- Look at the constraint types in the primal. Use results **5, 6** and **7** to write down the variable types in the dual.

- Look at the variable types in the primal. Use results **8, 9** and **10** to write down the constraint types in the dual.

Note

For a maximisation primal Table 2 must be read from right to left. This operation is justified by Theorem 1 given later.

Example 7.2
Use the table method to find the dual of the following linear programming problem :

$$\text{Minimise} \quad z = x_1 - 2x_2 + 3x_3$$
$$\text{Subject to :} \quad x_1 + x_2 + x_3 = 7 \quad \text{---- (1)}$$
$$5x_1 - 4x_2 + 6x_3 \geq 5 \quad \text{---- (2)}$$
$$x_1, x_2 \geq 0, \; x_3 \text{ unconstrained}$$

Solution
Using results 1 and 2 the dual objective function is :

$$\text{Maximise} \quad v = 7y_1 + 5y_2$$

Using result 4 the left-hand sides of the dual constraints are :

$$y_1 + 5y_2$$
$$y_1 - 4y_2$$
$$y_1 + 6y_2$$

Using result 3 the right-hand sides of the dual constraints are :

$$y_1 + 5y_2 \quad 1 \quad \text{---- (3)}$$
$$y_1 - 4y_2 \quad -2 \quad \text{---- (4)}$$
$$y_1 + 6y_2 \quad 3 \quad \text{---- (5)}$$

Constraint (1) in the primal corresponds to variable y_1 in the dual. Using result 6 y_1 is an unconstrained variable.

Constraint (2) in the primal corresponds to variable y_2 in the dual. Using result 7 y_2 is a non-negative variable.

Variable x_1 in the primal corresponds to constraint (3) in the dual. Using result 8 constraint (3) is a type 1 constraint.

Variable x_2 in the primal corresponds to constraint (4) in the dual. Using result 8 constraint (4) is a type 1 constraint.

Variable x_3 in the primal corresponds to constraint (5) in the dual. Using result 10 constraint (5) is a type 2 constraint.

Combining this information the dual linear programming problem becomes :

$$\begin{aligned}
\text{Maximise} \quad & v = 7y_1 + 5y_2 \\
\text{Subject to:} \quad & y_1 + 5y_2 \leq 1 \quad \text{---- (3)} \\
& y_1 - 4y_2 \leq -2 \quad \text{---- (4)} \\
& y_1 + 6y_2 = 3 \quad \text{---- (5)} \\
& y_1 \text{ unconstrained}, \; y_2 \geq 0
\end{aligned}$$

Multiplying through constraint (4) by -1 to make the right-hand side positive :

$$\begin{aligned}
\text{Maximise} \quad & v = 7y_1 + 5y_2 \\
\text{Subject to:} \quad & y_1 + 5y_2 \leq 1 \\
& -y_1 + 4y_2 \geq 2 \\
& y_1 + 6y_2 = 3 \\
& y_1 \text{ unconstrained}, \; y_2 \geq 0
\end{aligned}$$

7.4 Duality Theorems

The theorems given below are those required later in this chapter for a solving a primal linear programming problem using the solution of its dual and in Chapter 9 for performing a post-optimal analysis.

Theorem 1

The dual of the dual is the primal.

Theorem 2

(i)   If the primal has an optimal solution, the dual has an optimal solution.

(ii)   If the primal is feasible, the dual is bounded (and vice-versa).

(iii) If the primal is unbounded, the dual is infeasible (and vice-versa).

(iv)   If the primal is degenerate, the dual has a non-unique solution.

Theorem 3

(i)   If the ith dual constraint is **slack** at the optimum (*i.e.* inequality holds), the ith primal variable is zero.

(ii)   If the ith dual variable is non-zero at the optimum, the ith primal constraint is **tight** (*i.e.* equality holds).

Theorem 4

(i)   The coefficients of the **slack** variables in the objective function row of an optimal dual tableau are the <u>negatives</u> of the optimal values of the corresponding primal variables.

(ii)   The coefficients of the **surplus** variables in the objective function row of an optimal dual tableau are the optimal values of the corresponding primal variables.

Note

In a maximisation problem the coefficients in the objective function are multiplied by -1 when the problem is put into canonical form. Hence, in this case the coefficients of the slack variables in the objective function row of an optimal dual tableau are the optimal values of the corresponding primal variables and the coefficients of the surplus variables in this row are the <u>negatives</u> of the optimal values of the corresponding primal variables.

Theorem 5
Suppose that :

$\underline{c}_B^*$ contains the objective function coefficients from the <u>dual</u> corresponding to the basic variables in the optimal dual solution.

B^* contains the columns of the matrix A from the <u>dual</u> corresponding to the basic variables in the optimal dual solution.

Then :

- If the dual is a minimisation problem, the optimal solution of the primal is $\underline{x}^T = \underline{c}_B^* B^{*-1}$

- If the dual is a maximisation problem, the optimal solution of the primal is $\underline{x}^T = -\underline{c}_B^* B^{*-1}$

Theorem 6
The rate of change of the optimal standard primal objective function (*i.e.* z^*) with respect to the *i*th requirement (*i.e.* b_i) is the optimal value of the *i*th dual variable (*i.e.* y_i^*). That is :

$$\frac{\partial z^*}{\partial b_i} = y_i^*$$

7.5 Solving a Linear Programming Problem Using the Solution of the Dual Problem
On some occasions it is easier to form and solve the dual of a linear programming problem than it is to solve the primal. For example, the primal may contain type 2 and/or type 3 constraints and require a two-phase procedure to find its solution. Depending upon the variable types in the primal, the dual may contain type 1 constraints only and require a one-phase procedure to find its solution. Having solved the dual the solution of the primal can be found in three ways depending on the information available.

Case 1 : Given the Optimal Solution of the Dual
Suppose that the optimal solution of the dual is available. Then, the optimal solution of the primal can be found using Theorem 3. The procedure is :

- Substitute the optimal solution of the dual into the dual constraints. For each constraint that is **slack** (*i.e.* for which inequality holds), the corresponding primal variable is zero.

- Look at the optimal solution of the dual. For each variable that is non-zero the corresponding primal constraint is **tight** (*i.e.* equality holds). Use this information to write down a system of linear equations that can be solved to give the values of the other primal variables.

Example 7.3

Consider the following linear programming problem :

$$\text{Minimise} \quad z = 1000x_1 + 1000x_2 + 1200x_3$$
$$\text{Subject to :} \quad 4x_1 + 5x_2 + 3x_3 \geq 5 \quad \text{---- (1)}$$
$$5x_1 + 2x_2 + 8x_3 \geq 3 \quad \text{---- (2)}$$
$$x_1, x_2, x_3 \geq 0$$

The dual of this problem is :

$$\text{Maximise} \quad v = 5y_1 + 3y_2$$
$$\text{Subject to :} \quad 4y_1 + 5y_2 \leq 1000 \quad \text{---- (3)}$$
$$5y_1 + 2y_2 \leq 1000 \quad \text{---- (4)}$$
$$3y_1 + 8y_2 \leq 1200 \quad \text{---- (5)}$$
$$y_1, y_2 \geq 0$$

The optimal solution of the dual is :

$$v = 18000/17, \quad y_1 = 3000/17 \text{ and } y_2 = 1000/17$$

The relevant correspondences between the primal and the dual are :

Primal		Dual
x_1	$\leftrightarrow$	Constraint (3)
x_2	$\leftrightarrow$	Constraint (4)
x_3	$\leftrightarrow$	Constraint (5)
Constraint (1)	$\leftrightarrow$	y_1
Constraint (2)	$\leftrightarrow$	y_2

Table 3

Substituting the optimal dual solution into the dual constraints :

$(3) \Rightarrow 4(3000/17) + 5(1000/17) = 17000/17 = 1000$ *i.e.* this constraint is **tight**. Hence $x_1 \neq 0$

$(4) \Rightarrow 5(3000/17) + 2(1000/17) = 17000/17 = 1000$ *i.e.* this constraint is **tight**. Hence $x_2 \neq 0$

$(5) \Rightarrow 3(3000/17) + 8(1000/17) = 17000/17 = 1000$ *i.e.* this constraint is **slack**. Hence $x_3 = 0$

Since $y_1 > 0$, constraint (1) is tight. That is :

$$4x_1 + 5x_2 + 3(0) = 5 \quad \textit{i.e.} \quad 4x_1 + 5x_2 = 5 \quad \text{---- (6)}$$

Since $y_2 > 0$, constraint (2) is also tight. That is :

$$5x_1 + 2x_2 + 8(0) = 3 \quad i.e. \quad 5x_1 + 2x_2 = 3 \quad \text{---- (7)}$$

Solving equations (6) and (7) simultaneously $x_1 = 5/17$ and $x_2 = 13/17$. Hence, the optimal solution of the primal is :

$$z = 18000/17, \ x_1 = 5/17, \ x_2 = 13/17 \text{ and } x_3 = 0$$

Case 2 : Given the Optimal Dual Tableau

In this case the optimal solution of the primal can be found by reading off the optimal solution of the dual from the tableau and then using the procedure described in case 1. An alternative method is to use Theorem 4. The procedure here is to read off the coefficients of the slack and surplus variables from the objective function row of the optimal dual tableau. Then :

- The coefficients of the **slack** variables in the objective function row of the optimal dual tableau are the <u>negatives</u> of the optimal values of the corresponding primal variables.

- The coefficients of the **surplus** variables in the objective function row of the optimal dual tableau are the optimal values of the corresponding primal variables.

Example 7.4

Consider the following linear programming problem :

$$\begin{aligned}
\text{Maximise} \quad & z = 2x_1 + 6x_2 + 3/2 x_3 \\
\text{Subject to :} \quad & 1/2 x_1 + x_2 + 1/2 x_3 \leq 2 \\
& x_1 - x_2 - x_3 \leq 2 \\
& x_1, x_2 \geq 0, x_3 \leq 0
\end{aligned}$$

The dual of this problem is :

$$\begin{aligned}
\text{Minimise} \quad & v = 2y_1 + 2y_2 \\
\text{Subject to :} \quad & 1/2 y_1 + y_2 \geq 2 \quad \text{---- (1)} \\
& y_1 - y_2 \geq 6 \quad \text{---- (2)} \\
& 1/2 y_1 - y_2 \leq 3/2 \quad \text{---- (3)} \\
& y_1, y_2 \geq 0
\end{aligned}$$

Introducing surplus variables y_3, y_4, a slack variable y_5 and artificial variables a_1, a_2 this problem can be written in canonical form as :

$$\begin{aligned}
\text{Minimise} \quad & -v + 2y_1 + 2y_2 = 0 \\
& -w - 3/2 y_1 + y_3 + y_4 = -8 \\
\text{Subject to :} \quad & 1/2 y_1 + y_2 - y_3 + a_1 = 2 \\
& y_1 - y_2 - y_4 + a_2 = 6 \\
& 1/2 y_1 - y_2 + y_5 = 3/2 \\
& y_1, y_2, y_3, y_4, y_5, a_1, a_2 \geq 0
\end{aligned}$$

113

The optimal dual tableau (excluding the w-row) is :

$-z$	y_1	y_2	y_3	y_4	y_5	RHS
1	0	0	0	6	8	-24
0	0	1	0	-1	-2	3
0	0	0	1	-2	-3	11/2
0	1	0	0	-2	-2	9

The relevant correspondences between the dual and the primal are :

Dual	Slack Variable	Surplus Variable		Primal
Constraint (1)	-	y_3	↔	x_1
Constraint (2)	-	y_4	↔	x_2
Constraint (3)	y_5	-	↔	x_3

Table 4

From the optimal dual tableau :

- The objective function coefficient of the **surplus** variable y_3 is 0 i.e. $x_1 = 0$.

- The objective function coefficient of the **surplus** variable y_4 is 6 i.e. $x_2 = 6$.

- The objective function coefficient of the **slack** variable y_5 is 8 i.e. $x_3 = -8$.

Hence, the optimal solution of the primal is :

$$z = 24, \ x_1 = 0, \ x_2 = 6 \text{ and } x_3 = -8$$

Case 3 : Given the Optimal Transformation Matrix
In this case the optimal solution of the primal can be found using Theorem 5 i.e.

- If the dual is a minimisation problem, the optimal solution of the primal is $\underline{x}^T = \underline{c}_B^* B^{*-1}$

- If the dual is a maximisation problem, the optimal solution of the primal is $\underline{x}^T = -\underline{c}_B^* B^{*-1}$

The vector $\underline{x}^T$ can be extracted from the first row of the optimal transformation matrix i.e.

$$\underline{x}^T / -\underline{x}^T$$

$$T = \begin{bmatrix} 1 & \boxed{-\underline{c}_B B^{-1}} \\ \underline{0} & B^{-1} \end{bmatrix}$$

114

Example 7.5
Consider again the linear programming problem :

$$\text{Minimise} \quad z = 1000x_1 + 1000x_2 + 1200x_3$$
$$\text{Subject to :} \quad 4x_1 + 5x_2 + 3x_3 \geq 5 \quad \text{---- (1)}$$
$$5x_1 + 2x_2 + 8x_3 \geq 3 \quad \text{--- (2)}$$
$$x_1, x_2, x_3 \geq 0$$

The dual of this problem is :

$$\text{Maximise} \quad v = 5y_1 + 3y_2$$
$$\text{Subject to :} \quad 4y_1 + 5y_2 \leq 1000 \quad \text{---- (3)}$$
$$5y_1 + 2y_2 \leq 1000 \quad \text{---- (4)}$$
$$3y_1 + 8y_2 \leq 1200 \quad \text{---- (5)}$$
$$y_1, y_2 \geq 0$$

The relevant correspondences between the dual and the primal are :

Dual		Primal
Constraint (3)	$\leftrightarrow$	x_1
Constraint (4)	$\leftrightarrow$	x_2
Constraint (5)	$\leftrightarrow$	x_3

Table 5

The optimal transformation matrix for the dual is :

$$T = \begin{bmatrix} 1 & 5/17 & 13/17 & 0 \\ 0 & 5/17 & -4/17 & 0 \\ 0 & -2/17 & 5/17 & 0 \\ 0 & 2 & 1 & 1 \end{bmatrix}$$

Since the dual is a maximisation problem, the optimal solution of the primal is $\underline{x}^T = -\underline{c}_B^* B^{*-1}$. From the first row of the T matrix :

$$-\underline{c}_B^* B^{*-1} = \begin{bmatrix} 5/17 & 13/17 & 0 \end{bmatrix}$$

Hence, the optimal values of the primal variables are :

$$x_1 = 5/17, \quad x_2 = 13/17 \quad \text{and} \quad x_3 = 0$$

and the optimal solution of the primal is :

$$z = 18000/17, \quad x_1 = 5/17, \quad x_2 = 13/17 \quad \text{and} \quad x_3 = 0$$

7.6 Exercises 7a

1. Find the dual of the following linear programming problem :

   Minimise $\quad z = 1/2\,x_1 + x_2$

   Subject to : $\quad 2x_1 - x_2 \geq 1$

   $\qquad\qquad -x_1 + 2x_2 \geq 11$

   $\qquad\qquad x_1 + x_2 \geq 14$

   $\qquad\qquad x_1, x_2 \geq 0$

2. (i) Find the dual of the dual linear programming problem from question 1. What do you notice ?

   (ii) Which linear programming problem is easier to solve, the primal problem from question 1. or its dual ? Give a reason for your answer.

   (iii) Solve the dual linear programming problem from question 1. using the simplex method.

   (iv) The optimal solution of the primal linear programming problem from question 1. is $z = 67/6$, $x_1 = 17/3$ and $x_2 = 25/3$. Compare this solution with the optimal tableau from 2(iii). What do you notice ?

3. Use the table method to find the dual of each of the following linear programming problems :

   (i) Minimise $\quad z = x_1 + x_2 - x_3$

   Subject to : $\quad 2x_1 - x_2 - x_3 = 1$

   $\qquad\qquad 3x_1 + x_2 - x_3 \geq 3$

   $\qquad\qquad -x_1 + 2x_2 + 2x_3 \leq 4$

   $\qquad\qquad x_1, x_2, x_3 \geq 0$

   (ii) Maximise $\quad z = 2x_1 - x_2$

   Subject to : $\quad x_1 + x_2 \leq 1/2$

   $\qquad\qquad x_1 + 2x_2 \geq 4$

   $\qquad\qquad -x_1 + 5x_2 \leq 5$

   $\qquad\qquad 2x_1 + x_2 \geq 3$

   $\qquad\qquad x_1 \leq 0, x_2 \geq 0$

4. Convert each of the linear programming problems from question 3. to standard primal form and then find the dual of each problem. You will need to make suitable changes of variable in each case to make the two solutions agree.

5. Solve the dual linear programming problem from question 3(ii) using the simplex method. What can you deduce about the solution of the primal problem ?

6. Consider the following dual linear programming problem :

Minimise $\quad v = 2y_1 - 3y_2 - 5y_3$

Subject to : $\quad 2y_1 - y_2 + 4y_3 \leq 10$

$\qquad\qquad 3y_1 + y_2 - 3y_3 \leq 12$

$\qquad\qquad -y_1 - y_2 + 7y_3 \leq 8$

$\qquad\qquad y_1, y_2, y_3 \geq 0$

The optimal solution of this problem is $y_1 = 0$, $y_2 = 27$, $y_3 = 5$ and $v = -106$. Use this solution to find the optimal solution of the primal.

7. Consider the following dual linear programming problem :

Minimise $\quad v = -2y_1 - 2y_2$

Subject to : $\quad -y_1 + y_2 \leq 2$

$\qquad\qquad y_1 + y_2 \leq 6$

$\qquad\qquad 1/2\, y_1 - y_2 \leq 3/2$

$\qquad\qquad y_1, y_2 \geq 0$

The optimal simplex tableau for this problem is :

$-z$	y_1	y_2	y_3	y_4	y_5	RHS
1	0	0	0	2	0	12
0	0	0	1	1/3	4/3	6
0	0	1	0	1/3	-2/3	1
0	1	0	0	2/3	2/3	5

Use this tableau to find the optimal solution of the primal.

8. Consider the following dual linear programming problem :

Maximise $\quad v = y_1 + y_2 - 3/2$

Subject to : $\quad y_1 - y_2 \geq -1/2$

$\qquad\qquad y_1 \leq 3$

$\qquad\qquad y_1, y_2 \geq 0$

The optimal transformation matrix for this problem is :

$$T = \begin{bmatrix} 1 & -1 & 2 \\ 0 & 1 & 1 \\ 0 & 0 & 1 \end{bmatrix}$$

Use this matrix to find the optimal solution of the primal.

7.7 The Dual Simplex Method

Suppose that the vector $\bar{b}$ contains the values on the right-hand sides of the constraints in the current simplex tableau and that the vector $\bar{c}$ contains the objective function coefficients in the current simplex tableau. The simplex method starts with a basic **feasible** solution (*i.e.* one in which $\underline{b} \geq \underline{0}$) and then proceeds to find an **optimal** solution (*i.e.* one in which $\bar{c} \geq \underline{0}$). It is often necessary to solve linear programming problems in which the initial solution is optimal but infeasible. Problems like this arise when :

- Finding a primal solution by forming and solving the dual problem.

- Solving problems in which the objective function coefficients are positive costs *e.g.* blending or mixing problems.

- Modelling or solving integer constrained linear programming problems *i.e.* when additional constraints must be added to the linear programming problem after the optimal solution of the corresponding non-integer constrained problem has been found.

One way of solving these problems is to interchange the roles of feasibility and optimality *i.e.* to use an algorithm that starts with an **optimal** solution (*i.e.* one in which $\underline{c} \geq \underline{0}$) and then proceeds to find a **feasible** solution (*i.e.* one in which $\bar{b} \geq \underline{0}$). This leads to the idea of a **dual simplex method**.

Development of the Algorithm

The main steps in the simplex method are :

1. Start with a basic <u>feasible</u> solution (*i.e.* one in which $\underline{b} \geq \underline{0}$).

2. Check for <u>optimality</u> *i.e.* is $\bar{c} \geq \underline{0}$? If yes then **stop**.

3. Choose the pivot <u>column</u> to be the largest negative element of the vector $\bar{c}$.

4. Choose the pivot <u>row</u> using a rule that ensures that <u>feasibility</u> is maintained.

5. Perform Jordan elimination using the chosen element as pivot.

6. Go back to Step 2.

Dualising these steps, the algorithm for the dual simplex method is :

1. Start with an <u>optimal</u> solution (*i.e.* one in which $\underline{c} \geq \underline{0}$).

2. Check for <u>feasibility</u> *i.e.* is $\bar{b} \geq \underline{0}$? If yes then **stop**.

3. Choose the pivot <u>row</u> to be the largest negative element of the vector $\bar{b}$.

4. Choose the pivot <u>column</u> using a rule that ensures that <u>optimality</u> is maintained.

5. Perform Jordan elimination using the chosen element as pivot.

6. Go back to Step 2.

The Pivot Selection Rule

In the simplex method the pivot selection rule involves dividing the _b-coefficients_ on the right-hand sides of the constraints by the _positive coefficients_ in the _pivot column._ Dualising this, the pivot selection rule in the dual simplex method must involve dividing the _c-coefficients_ in the objective function by the _negative coefficients_ in the _pivot row._ The pivot selection rule in the simplex method produces positive ratios. For the pivot selection rule for the dual simplex method to produce positive ratios the negatives of the negative coefficients in the pivot row must be used.

Consider the following optimal but infeasible tableau :

$-z$	x_1	x_2	x_3	x_4	RHS
1	5	3	1	2	20
.	.	.	.	.	.
.	.	.	.	.	.
0	4	-6	-1	-1	-10
.	.	.	.	.	.

Row p points to the row `0  4  -6  -1  -1  -10`. Largest negative coefficient points to the x_2 column.

The choices for the pivot column are :

a. The x_2 column

In this case $\dfrac{c_2}{-a_{p2}} = \dfrac{3}{-(-6)} = \dfrac{1}{2}$.

Performing Jordan elimination using the **-6** in the x_2 column as pivot produces the following tableau:

$-z$	x_1	x_2	x_3	x_4	RHS
1	7	0	1/2	3/2	15
.	.	.	.	.	.
.	.	.	.	.	.
0	$-2/3$	1	1/6	1/6	5/3
.	.	.	.	.	.

This option produces feasibility and maintains optimality.

b. The x_3 column

In this case $\dfrac{c_3}{-a_{p3}} = \dfrac{1}{-(-1)} = 1.$

Performing Jordan elimination using the **-1** in the x_3 column as pivot produces the following tableau:

$-z$	x_1	x_2	x_3	x_4	RHS
1	9	-3	0	1	10
.	.	.	.	.	.
.	.	.	.	.	.
0	-4	6	1	1	10
.	.	.	.	.	.

This option produces feasibility but does **not** maintain optimality.

c. The x_4 column

In this case $\dfrac{c_4}{-a_{p4}} = \dfrac{2}{-(-1)} = 2$.

Performing Jordan elimination using the **-1** in the x_4 column as pivot produces the following tableau:

$-z$	x_1	x_2	x_3	x_4	RHS
1	13	-9	-1	0	0
.	.	.	.	.	.
0	-4	6	1	1	10
.	.	.	.	.	.

This option produces feasibility but does **not** maintain optimality.

Hence, to produce feasibility and maintain optimality option **a** must be chosen. Notice that this option is the one that gives the smallest value of $\dfrac{c_k}{-a_{pk}}$. Hence, the pivot selection rule in the dual simplex method is :

$$\frac{c_k}{-a_{pk}} = \min\left\{\frac{c_j}{-a_{pj}} \;\middle|\; a_{pj} < 0 \;;\; 1 \le j \le n\right\}$$

where n is the number of structural variables.

Summary of the Dual Simplex Method
The initial solution of the linear programming problem must be optimal *i.e.* $\underline{c} \ge \underline{0}$.

Step 1
Write the linear programming problem in **canonical form**. Since the b-coefficients on the right-hand sides of the constraints can be negative this can be done <u>without</u> using artificial variables. Each type 2 constraint can be replaced by a type 1 constraint and a type 3 constraint. The surplus variable in each type 3 constraint can be made basic by multiplying through the constraint by -1.

Step 2
Examine the b-coefficients on the right-hand sides of the constraints. If $\bar{b}_i \ge 0 \;\forall i$ then **stop**, the optimal and feasible solution of the problem has been found.

Step 3
Select the pivot row. This is the row that has the largest negative b-coefficient on the right-hand side. Suppose that the pivot row is **row p.**

Step 4
Select the pivot column using the **pivot selection rule** $\dfrac{c_k}{-a_{pk}} = \min\left\{\dfrac{c_j}{-a_{pj}} \;\middle|\; a_{pj} < 0 \;;\; 1 \le j \le n\right\}$

Step 5

Perform **Jordan elimination** using a_{pk} as pivot.

Step 6

Go back to **Step 2**.

Problem Cases

When using the dual simplex method :

- A **non-unique solution** can be identified and dealt with in the same way as it would be when using the simplex method and the two-phase simplex method.

- A linear programming problem is **infeasible** if feasibility cannot be restored *i.e.* if at any stage during the solution the pivot selection rule breaks down.

- A linear programming problem **cannot be unbounded**. Since the problem is optimal *i.e.* $\underline{c} \geq \underline{0}$ and the structural variables are non-negative *i.e.* $x_i \geq 0 \; \forall i$, the optimal value of z *i.e.*

$$z_{min} = c_0 + \sum_{i=1}^{n} c_i x_i$$

is always bounded below by c_0.

- **Degeneracy** has no affect (*i.e.* does not cause cycling). This is because the rows with zeros on the right-hand side are never selected to be the pivot row when applying the method to a non-optimal tableau.

Example 7.6

Consider the following linear programming problem :

$$\text{Minimise} \quad z = 2x_1 + 3x_2$$
$$\text{Subject to :} \quad x_1 + x_2 = 20 \quad \text{---- (1)}$$
$$x_1 - x_2 \leq 5 \quad \text{---- (2)}$$
$$2x_1 + 3x_2 \geq 25 \quad \text{---- (3)}$$
$$x_1, x_2 \geq 0$$

The objective function can be written as :

$$-z + 2x_1 + 3x_2 = 0$$

Constraint (1) can be replaced by the pair of constraints :

$$x_1 + x_2 \leq 20 \quad \text{---- (4)}$$
$$x_1 + x_2 \geq 20 \quad \text{---- (5)}$$

Adding a slack variable x_3 to constraint (4) and subtracting a surplus variable x_4 from constraint (5) :

$$x_1 + x_2 + x_3 = 20$$

$$x_1 + x_2 - x_4 = 20 \quad \text{---- (6)}$$

Multiplying through constraint (6) by -1 to make x_4 a basic variable :

$$-x_1 - x_2 + x_4 = -20$$

Adding a slack variable x_5 to constraint (2) :

$$x_1 - x_2 + x_5 = 5$$

Subtracting a surplus variable x_6 from constraint (3) :

$$2x_1 + 3x_2 - x_6 = 25 \quad \text{---- (7)}$$

Multiplying through constraint (7) by -1 to make x_6 a basic variable :

$$-2x_1 - 3x_2 + x_6 = -25$$

Combining these expressions the linear programming problem can be written in canonical form as :

$$
\begin{aligned}
\text{Minimise} \quad & -z + 2x_1 + 3x_2 = 0 \\
\text{Subject to :} \quad & x_1 + x_2 + x_3 = 20 \\
& -x_1 - x_2 + x_4 = -20 \\
& x_1 - x_2 + x_5 = 5 \\
& -2x_1 - 3x_2 + x_6 = -25 \\
& x_1, x_2, x_3, x_4, x_5, x_6 \geq 0
\end{aligned}
$$

The initial solution of this problem is optimal but infeasible.

The initial tableau is :

Pivot Column

$-z$	x_1	x_2	x_3	x_4	x_5	x_6	**RHS**
1	2	3	0	0	0	0	0
0	1	1	1	0	0	0	20
0	-1	-1	0	1	0	0	-20
0	1	-1	0	0	1	0	5
0	-2	-3	0	0	0	1	-25

Pivot Row   (Joint) Smallest Ratio   Largest Negative Coefficient

122

Using the pivot selection rule :

$$S = \left\{ \frac{2}{-(-2)}, \frac{3}{-(-3)} \right\} \quad i.e. \quad S = \{1,1\}$$

The smallest ratio is 1. Taking the first occurrence, the pivot is the -2 in the x_1 column. Performing Jordan elimination and then continuing, the dual simplex tableau is :

$-z$	x_1	x_2	x_3	x_4	x_5	x_6	RHS
1	0	0	0	0	0	1	-25
0	0	$-1/2$	1	0	0	$1/2$	$15/2$
0	0	$1/2$	0	1	0	$-1/2$	$-15/2$
0	0	$-5/2$	0	0	1	$1/2$	$-15/2$
0	1	$3/2$	0	0	0	$-1/2$	$25/2$
1	0	1	0	2	0	0	-40
0	0	0	1	1	0	0	0
0	0	-1	0	-2	0	1	15
0	0	-2	0	1	1	0	-15
0	1	1	0	-1	0	0	20
1	0	0	0	$5/2$	$1/2$	0	$-95/2$
0	0	0	1	1	0	0	0
0	0	0	0	$-5/2$	$-1/2$	1	$45/2$
0	0	1	0	$-1/2$	$-1/2$	0	$15/2$
0	1	0	0	$-1/2$	$1/2$	0	$25/2$

Since all of the coefficients in the z-row are greater than or equal to zero and all of the b-coefficients on the right-hand sides of the constraints are greater than or equal to zero, the solution given in this tableau is now **optimal** and **feasible**. The solution of the linear programming problem is :

$$z = 95/2, \ x_1 = 25/2, \ x_2 = 15/2, \ x_3 = 0, \ x_4 = 0, \ x_5 = 0 \text{ and } x_6 = 45/2$$

7.8 The Artificial Constraint Technique

The dual simplex method requires the initial solution of the linear programming problem to be optimal (*i.e.* $\underline{c} \geq \underline{0}$). If it is not optimal then optimality can sometimes be achieved by adding an **artificial constraint** to the problem in the form :

$$\sum_{j=1}^{n} x_j \leq M$$

where the x_j are the structural variables and M is a large positive quantity that bounds the x_j above so that the artificial constraint does not impose any further restrictions on the variables *i.e.* so that the solution of the augmented problem is the same as the solution of the original problem.

The Procedure
To solve a linear programming problem using the artificial constraint technique :

- Write the augmented problem in canonical form. To do this add a slack variable x_s to the artificial constraint so that it can be written in the form :

$$\sum_{j=1}^{n} x_j + x_s = M$$

- Choose the artificial constraint row to be the pivot row. Suppose that the pivot row is **row p.**

- Choose the pivot column to be **column k** where x_k is the variable that has the largest negative coefficient in the objective function.

- Perform Jordan elimination using the element a_{pk} as pivot to produce an optimal tableau.

- Apply the dual simplex method to this tableau (if necessary) to obtain a feasible solution.

Outcomes
The possible outcomes of the artificial constraint technique are :

1) The procedure is <u>unable</u> to find a feasible solution. In this case the original linear programming problem is **infeasible**.

2) The procedure is <u>able</u> to find a feasible solution. In this case the value of the slack variable in the artificial constraint (*i.e.* x_s) must be examined.

- If $x_s > 0$ then the optimal solution of the augmented problem is the **optimal solution** of the original linear programming problem.

- If $x_s = 0$ then there are two possibilities *i.e.*

  - If the optimal value of z (*i.e.* z^*) depends upon M *i.e.* $z^* = f(M)$ then the original linear programming problem is **unbounded**.

  - If the optimal value of z is independent of M *i.e.* $z^* \neq f(M)$ then the original linear programming problem has a **non-unique solution**.

Example 7.7
Consider the following linear programming problem :

$$
\begin{aligned}
\text{Maximise} \quad & z = 2x_1 - 4x_2 \\
\text{Subject to :} \quad & 4x_1 + 2x_2 \geq 8 \quad \text{---- (1)} \\
& -x_1 + x_2 \geq 10 \quad \text{---- (2)} \\
& x_1, x_2 \geq 0
\end{aligned}
$$

The initial solution of this problem is feasible and hence the optimal solution can be found using the two-phase simplex method. However, it is sometimes easier to solve problems of this kind using the artificial constraint technique (since it is a one-phase procedure).

Converting to a minimisation problem, the objective function can be written as :

$$-\bar{z} - 2x_1 + 4x_2 = 0$$

Subtracting a surplus variable x_3 from constraint (1) :

$$4x_1 + 2x_2 - x_3 = 8$$

Multiplying through this constraint by -1 to make x_3 a basic variable :

$$-4x_1 - 2x_2 + x_3 = -8$$

Subtracting a surplus variable x_4 from constraint (2) :

$$-x_1 + x_2 - x_4 = 10$$

Multiplying through this constraint by -1 to make x_4 a basic variable :

$$x_1 - x_2 + x_4 = -10$$

Combining these expressions, the linear programming problem can be written in canonical form as :

Minimise $\quad -\bar{z} - 2x_1 + 4x_2 = 0$

Subject to : $\quad -4x_1 - 2x_2 + x_3 = -8$

$\quad\quad\quad\quad x_1 - x_2 + x_4 = -10$

$\quad\quad\quad\quad x_1, x_2, x_3, x_4 \geq 0$

The initial solution of this linear programming problem is neither optimal nor feasible. The artificial constraint is :

$$x_1 + x_2 \leq M$$

Adding a slack variable x_s :

$$x_1 + x_2 + x_s = M$$

Combining this with the other expressions the augmented problem becomes :

Minimise $\quad -\bar{z} - 2x_1 + 4x_2 = 0$

Subject to : $\quad -4x_1 - 2x_2 + x_3 = -8$

$\quad\quad\quad\quad x_1 - x_2 + x_4 = -10$

$\quad\quad\quad\quad x_1 + x_2 + x_s = M$

$\quad\quad\quad\quad x_1, x_2, x_3, x_4, x_s \geq 0$

The initial tableau is :

── Largest Negative Coefficient

$-\bar{z}$	x_1	x_2	x_3	x_4	x_s	RHS
1	-2 ← 4	4	0	0	0	0
0	-4	-2	1	0	0	-8
0	1	-1	0	1	0	-10
0	1	1	0	0	1	M

Pivot Column ────┘   Pivot Row

Pivoting on the greyed element in the x_1 column :

── Pivot Column

$-\bar{z}$	x_1	x_2	x_3	x_4	x_s	RHS
1	0	6	0	0	2	$2M$
0	0	2	1	0	4	$-8+4M$
0	0	-2	0	1	-1	$-10-M$
0	1	1	0	0	1	M

── Pivot Row   Smallest Ratio   Largest Negative Coefficient

The solution given in this tableau is now optimal but infeasible. Applying the dual simplex method to this tableau :

$-\bar{z}$	x_1	x_2	x_3	x_4	x_s	RHS
1	0	2	0	2	0	$2M$
0	0	-6	1	4	0	-48
0	0	2	0	-1	1	$10+M$
0	1	-1	0	1	0	-10
1	0	0	1/3	10/3	0	-36
0	0	1	-1/6	-2/3	0	8
0	0	0	1/3	1/3	1	$-6+M$
0	1	0	-1/6	1/3	0	-2
1	2	0	0	4	0	-40
0	-1	1	0	-1	0	10
0	2	0	0	1	1	$-10+M$
0	-6	0	1	-2	0	12

The solution given in this tableau is now optimal and feasible. The slack variable in the artificial constraint *i.e.* x_s is positive *i.e.* $x_s > 0$. Hence, the optimal solution of the augmented problem is the optimal solution of the original linear programming problem. From the tableau above the optimal solution is :

$$-\bar{z} = -40 \quad i.e. \quad z = -40, \quad x_1 = 0 \text{ and } x_2 = 10$$

Example 7.8
Consider the following linear programming problem :

$$\text{Minimise} \quad z = -x_1 + x_2 - x_3 - 7$$
$$\text{Subject to :} \quad x_1 + x_2 - x_3 \geq 1 \quad \text{---- (1)}$$
$$-x_1 + x_2 + 2x_3 \geq 2 \quad \text{---- (2)}$$
$$x_1, x_2, x_3 \geq 0$$

The objective function can be written as :

$$-z - x_1 + x_2 - x_3 = 7$$

Subtracting a surplus variable x_4 from constraint (1) :

$$x_1 + x_2 - x_3 - x_4 = 1$$

Multiplying through this constraint by -1 to make x_4 a basic variable :

$$-x_1 - x_2 + x_3 + x_4 = -1$$

Subtracting a surplus variable x_5 from constraint (2) :

$$-x_1 + x_2 + 2x_3 - x_5 = 2$$

Multiplying through this constraint by -1 to make x_5 a basic variable :

$$x_1 - x_2 - 2x_3 + x_5 = -2$$

Combining these expressions, the linear programming problem can be written in canonical form as :

$$\text{Minimise} \quad -z - x_1 + x_2 - x_3 = 7$$
$$\text{Subject to :} \quad -x_1 - x_2 + x_3 + x_4 = -1$$
$$x_1 - x_2 - 2x_3 + x_5 = -2$$
$$x_1, x_2, x_3, x_4, x_5 \geq 0$$

The initial solution of this linear programming problem is neither optimal nor feasible. The artificial constraint is :

$$x_1 + x_2 + x_3 \leq M$$

Adding a slack variable x_s :

$$x_1 + x_2 + x_3 + x_s = M$$

Combining this with the other expressions the augmented problem becomes :

$$\text{Minimise} \quad -z - x_1 + x_2 - x_3 = 7$$
$$\text{Subject to :} \quad -x_1 - x_2 + x_3 + x_4 = -1$$
$$x_1 - x_2 - 2x_3 + x_5 = -2$$
$$x_1 + x_2 + x_3 + x_s = M$$
$$x_1, x_2, x_3, x_4, x_5, x_s \geq 0$$

The initial tableau is :

——— (Joint) Largest Negative Coefficient

$-z$	x_1	x_2	x_3	x_4	x_5	x_s	RHS
1	-1	1	-1	0	0	0	7
0	-1	-1	1	1	0	0	-1
0	1	-1	-2	0	1	0	-2
0	1	1	1	1	0	0	1	M

Pivot Column ———   Pivot Row

Pivoting on the greyed element in the x_1 column :

——— Pivot Column

$-z$	x_1	x_2	x_3	x_4	x_5	x_s	RHS
1	0	2	0	0	0	1	$M+7$
0	0	0	2	1	0	1	$-1+M$
0	0	-2	-3	0	1	-1	$-M-2$
0	1	1	1	0	0	1	M

Pivot Row   Smallest Ratio   Largest Negative Coefficient ———

The solution given in this tableau is now optimal but infeasible. Applying the dual simplex method to this tableau using the greyed element in the x_3 column as pivot :

$-z$	x_1	x_2	x_3	x_4	x_5	x_s	RHS
1	0	2	0	0	0	1	$M+7$
0	0	$-4/3$	0	1	$2/3$	$1/3$	$M/3-7/3$
0	0	$2/3$	1	0	$-1/3$	$1/3$	$M/3+2/3$
0	1	$1/3$	0	0	$1/3$	$2/3$	$2M/3-2/3$

The solution given in this tableau is now optimal and feasible. However, the slack variable in the artificial constraint *i.e.* x_s is non-basic *i.e.* $x_s = 0$. In addition, the optimal value of z (*i.e.* z^*) depends upon M *i.e.* $z^* = f(M)$. Hence, the original linear programming problem is **unbounded**.

7.8 Choice of Method

Linear programming problems can be solved using a variety of numerical algorithms *e.g.* the simplex methods, the revised simplex methods, the dual simplex method, the artificial constraint technique, *etc.* For a particular problem one of these methods will be the most computationally efficient *i.e.* require the fewest number of variables and the fewest number of calculations to find the optimal solution. Consider a linear programming problem with n structural variables and m structural constraints. Then :

- If the problem is not optimal and not feasible then it must be solved using the artificial constraint technique.

- If all of the constraints are type 3 then the dual simplex method is the most computationally efficient. This is a one-phase procedure and requires $n+2$ variables. The appropriate versions of the simplex and revised methods are two-phase procedures and require $n+2m$ variables.

- If $n \geq 4m$ then the revised simplex methods are the most computationally efficient.

- If $n < 4m$ then the simplex methods are the most computationally efficient.

7.9 Exercises 7b

1. Use the dual simplex method to solve the following linear programming problems :

   (i) Minimise $z = 4x_1 + 8x_2 + 3x_3$
        Subject to : $x_1 + x_2 \geq 2$
                       $2x_2 + x_3 \geq 5$
                       $x_1, x_2, x_3 \geq 0$

   (ii) Maximise $z = -x_1 - x_2$
        Subject to : $2x_1 + x_2 + x_3 = 50$
                       $x_1 + x_2 \geq 30$
                       $-x_1 + 2x_2 \geq 25$
                       $x_1, x_2, x_3 \geq 0$

2. (i) Sketch the feasible region for the linear programming problem solved using the dual simplex method in Example 7.6 *i.e.*

        Minimise $z = 2x_1 + 3x_2$
        Subject to : $x_1 + x_2 = 20$
                       $x_1 - x_2 \leq 5$
                       $2x_1 + 3x_2 \geq 25$
                       $x_1, x_2 \geq 0$

   (ii) Annotate your diagram with the basic solutions produced by the dual simplex algorithm. What do you notice about these solutions ?

3. A metal recycling company processes three grades of scrap to produce iron and nickel for the aircraft industry. The table below shows the amount (as a %) of iron and nickel recoverable from each tonne of scrap :

	Grade 1	Grade 2	Grade 3
Iron	30	40	20
Nickel	20	30	40

The purchase costs of the scrap are £50, £60 and £40 for each tonne of Grade 1, Grade 2 and Grade 3 respectively. The company receives an order for 3000 tonnes of iron and 2500 tonnes of nickel.

(i) Formulate a linear programming model from which the company can determine how many tonnes of each grade of scrap to purchase in order to satisfy this order at minimum cost.

(ii) Solve your linear programming model from 3(i) using the dual simplex method.

4. Use the artificial constraint technique to solve the following linear programming problems :

(i) Minimise $z = 2x_1 - 2x_2 + 4x_3$

Subject to : $-x_1 + 2x_2 + 3x_3 \le 9$

$-x_2 + 4x_3 \ge 1$

$-x_1 + 5x_2 \le 1$

$x_1, x_2, x_3 \ge 0$

(ii) Minimise $z = -x_1 - 2x_2 - x_3$

Subject to : $-x_1 + 2x_2 + x_3 = 4$

$-2x_1 - x_3 \ge 3$

$3x_1 + 2x_2 - x_3 \le 8$

$x_1, x_2, x_3 \ge 0$

(iii) Maximise $z = 4x_1 - 2x_2 - 3$

Subject to : $-2x_1 + x_2 + x_3 \le 10$

$x_1 - x_2 + 2x_3 \le 8$

$x_1, x_2, x_3 \ge 0$

(iv) Maximise $z = x_1 + x_2$

Subject to : $x_2 \le 8$

$-x_1 + x_2 \ge -4$

$x_1 + x_2 \le 12$

$x_1, x_2 \ge 0$

(v) Minimise $z = -3x_1 - 4x_2 - 5x_3$

    Subject to : $x_1 + x_2 + x_3 \geq 4$

                 $2x_1 + 3x_2 + 4x_3 \leq 6$

                 $x_1, x_2, x_3 \geq 0$

5. State which method you would recommend for solving each of the following linear programming problems. Give reasons for your answers.

(i) Minimise $z = 2x_1 + 3x_2 - 5$

    Subject to : $4x_1 + 2x_2 \leq 20$

                 $2x_1 + x_2 \leq 30$

                 $x_1 - 6x_2 \leq 10$

                 $5x_1 + 7x_2 \leq 15$

                 $4x_1 - x_2 \leq 14$

                 $x_1, x_2 \geq 0$

(ii) Minimise $-z - 3x_1 + x_2 - 2x_3 = 5$

    Subject to : $-x_1 - 3x_2 + x_3 - 4x_4 = -1$

                 $x_1 + 5x_2 - 4x_3 - x_5 = -8$

                 $x_1, x_2, x_3, x_4, x_5 \geq 0$

(iii) Maximise $z = 2x_1 + 4x_2 - 7x_3 + 10$

    Subject to : $5x_1 + x_2 + 10x_3 \geq 50$

                 $4x_1 + 3x_2 - 6x_3 \geq 47$

                 $x_1, x_2, x_3 \geq 0$

(iv) Minimise $z = 2x_1 + 3x_2 - x_3 + 5x_4 - 2x_5 + x_6 - 4x_7 + 3x_8 - 6x_9$

    Subject to : $3x_1 + x_2 - 4x_3 + 6x_4 - x_5 - 2x_6 + 3x_7 + 2x_8 - x_9 \leq 80$

                 $11x_1 - 4x_2 - 2x_3 + x_4 + 2x_5 + x_6 - x_7 - x_8 + 5x_9 \leq 60$

                 $x_1, x_2, x_3, x_4, x_5, x_6, x_7 \geq 0$

8. Integer Linear Programming

8.1 Introduction

An **integer** linear programming problem is one in which integer constraints are imposed on some or all of the variables. Problems of this kind arise when the variables represent discrete and indivisible entities such as people, machines, components, *etc*. If all of the variables are constrained to be integer then the problem is called a **pure** integer linear programming problem. If only some of the variables are constrained to be integer then the problem is called a **mixed** integer linear programming problem.

Example 8.1

Consider the following pure integer linear programming problem :

$$\text{Maximise} \quad z = x_1 + x_2$$
$$\text{Subject to :} \quad 2x_2 \le 9$$
$$9x_1 + 4x_2 \le 54$$
$$x_1, x_2 \ge 0$$
$$z, x_1, x_2 \text{ integer}$$

Solving this problem using the graphical method, the optimal solutions of the integer constrained and non-integer constrained problems are as follows :

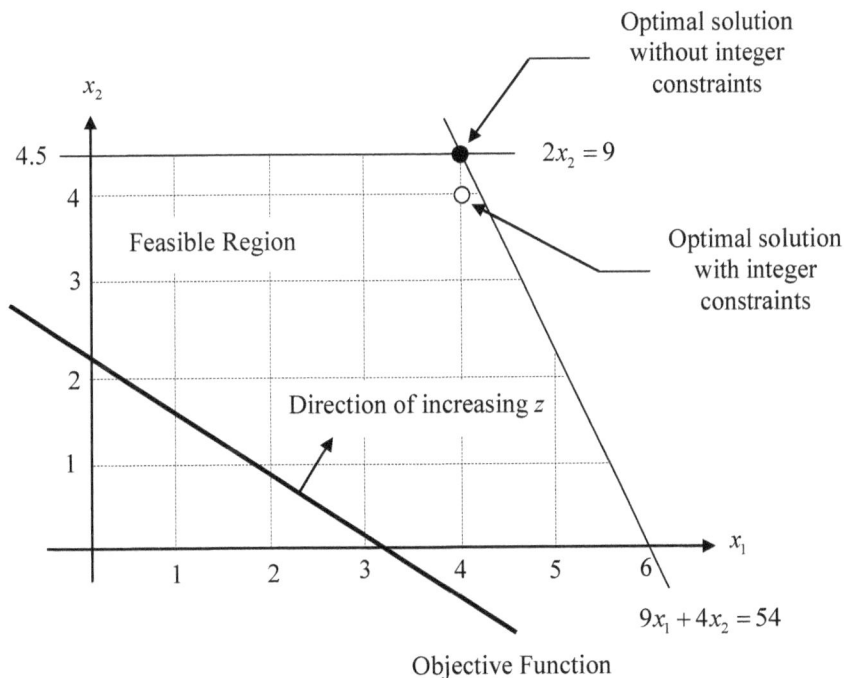

Objective Function

132

It can be seen from this diagram that the optimal solution of the integer constrained problem lies at one of the grid points inside the feasible region while the optimal solution of the non-integer constrained problem lies on the boundary of the feasible region.

Use of Rounding
One way to solve an integer linear programming problem is to solve the associated non-integer constrained problem and then to round the solutions obtained to the nearest integer values. However, this approach does not always work. For example, consider the linear programming problem from Example 8.1. Rounding the non-integer solution in this case produces the solution $x_1 = 4$, $x_2 = 5$ which is infeasible. On other occasions rounding a non-integer solution may produce an integer solution that is feasible but non-optimal *i.e.* it may not be the integer point that gives the maximum or minimum value of z. Hence, in general rounding non-integer solutions should be avoided.

Solving Integer Linear Programming Problems
The most commonly used methods for solving integer linear programming problems are :

- **Cutting Plane Methods**.

- **The Branch and Bound Method**.

8.2 Cutting Plane Methods
The first stage when using a cutting plane method is to solve the associated non-integer constrained problem. If all of the variables have integer values then the optimal solution has been found and no further work is required. However, if this is not the case then an additional constraint called a **cut** is added to the problem. The cut is designed to make the current solution **infeasible** *i.e.* to remove the part of the feasible region that contains the current solution but to ensure that the feasible region still contains the optimal integer solution. Once the cut has been added to the problem the dual simplex method is used to restore feasibility. This process is repeated until either the optimal integer solution is found or until it becomes clear that no optimal integer solution exists (*i.e.* until the pivot selection rule in the dual simplex method breaks down indicating that the problem is infeasible). A variety of cutting plane methods is available. Some can be used for solving pure integer linear programming problems only while others can be used for solving both pure and mixed integer linear programming problems. This chapter considers one of the simplest of these methods. This is called the **cyclic algorithm**.

The Cyclic Algorithm
The cyclic algorithm can be used for solving pure integer linear programming problems only. It requires that all entries in the vectors c and b and the matrix A are integer.

Notation
Any real number x can be written in the form :

$$x = [x] + F(x)$$

where $[x]$ is the **integer part** of x and $F(x)$ is the **fractional part** of x *i.e.* $0 \le F(x) < 1$.

Examples
- $5.6 = 5 + 0.6$ *i.e.* $[5.6] = 5$ and $F(5.6) = 0.6$
- $4.0 = 4 + 0.0$ *i.e.* $[4.0] = 4$ and $F(4.0) = 0.0$

- $-4.7 = -5 + 0.3$ *i.e.* $[-4.7] = -5$ and $F(-4.7) = 0.3$
- $-8.2 = -9 + 0.8$ *i.e.* $[-8.2] = -9$ and $F(-8.2) = 0.8$

etc.

The Cut

Consider the *i*th constraint in the optimal solution of the associated non-integer constrained problem *i.e.*

$$\sum_{j=1}^{n} a_{ij}x_j + x = b_i$$

where $x_1, x_2, \ldots, x_n$ are non-basic variables and x is the basic variable in that constraint. The fractional parts of each side of this constraint must be equal *i.e.*

$$F\left(\sum_{j=1}^{n} a_{ij}x_j + x\right) = F(b_i)$$

Writing a_{ij} in terms of its integer part and fractional part :

$$F\left(\sum_{j=1}^{n}\left([a_{ij}] + F(a_{ij})\right)x_j + x\right) = F(b_i)$$

$$F\left(\sum_{j=1}^{n}\left([a_{ij}]x_j + F(a_{ij})x_j\right) + x\right) = F(b_i)$$

$$F\left(\sum_{j=1}^{n}[a_{ij}]x_j + \sum_{j=1}^{n}F(a_{ij})x_j + x\right) = F(b_i)$$

i.e. $\quad F\left(\sum_{j=1}^{n}[a_{ij}]x_j\right) + F\left(\sum_{j=1}^{n}F(a_{ij})x_j\right) + F(x) = F(b_i)$

For any all integer solution :

$$F\left(\sum_{j=1}^{n}[a_{ij}]x_j\right) = 0 \text{ and } F(x) = 0$$

Hence :

$$F\left(\sum_{j=1}^{n}F(a_{ij})x_j\right) = F(b_i) \quad \text{---- (1)}$$

Expression (1) shows that the fractional part of $\sum\limits_{j=1}^{n} F(a_{ij})x_j$ is equal to the fractional part of b_i.

Since $\sum\limits_{j=1}^{n} F(a_{ij})x_j$ also has an integer part, it must be true that :

$$\sum_{j=1}^{n} F(a_{ij})x_j \geq F(b_i) \quad \text{---- (2)}$$

Suppose that this constraint is added to the associated non-integer constrained problem. Subtracting a surplus variable x_s :

$$\sum_{j=1}^{n} F(a_{ij})x_j - x_s = F(b_i)$$

Multiplying through by -1 to make this constraint canonical :

$$-\sum_{j=1}^{n} F(a_{ij})x_j + x_s = -F(b_i)$$

Since $F(b_i) \geq 0$ the right-hand side of this constraint is negative. Hence, adding this constraint has made the current solution infeasible *i.e.* the current solution is now outside of the feasible region.

In summary, adding constraint (2) to the associated non-integer constrained problem has removed the part of the feasible region that contains the current solution but has ensured that the feasible region still contains the optimal integer solution. Hence, constraint (2) is the required cut.

The Procedure
To solve a pure integer linear programming problem using the cyclic algorithm :

1. Solve the associated non-integer constrained problem. If all of the variables have integer values then the optimal solution has been found and no further work is required. Otherwise :

2. Read the tableau from top to bottom and find the first row whose right-hand side is not integer. The z-row can be selected. Use this row to generate the cut :

$$\sum_{j=1}^{n} F(a_{ij})x_j \geq F(b_i)$$

    Generating the cut in this way ensures that the algorithm terminates in a finite number of iterations provided that cycling does not occur and that a feasible solution of problem exists.

3. Subtract a surplus variable from the left-hand side and then multiply through by -1 to make the cut canonical.

135

4. Add the cut as an additional constraint to the previous optimal tableau.

5. Use the dual simplex method to restore feasibility.

6. If all of the variables have integer values then the optimal solution has been found and no further work is required. Otherwise go back to step 2.

Example 8.2
Use the cyclic algorithm to solve the following pure integer linear programming problem :

$$\begin{aligned}
\text{Minimise} \quad & z = -x_1 - 5x_2 \\
\text{Subject to :} \quad & -x_1 + x_2 \leq 1 \\
& 2x_1 + 3x_2 \leq 6 \\
& x_1, x_2 \geq 0 \\
& z, x_1, x_2 \text{ integer}
\end{aligned}$$

Solution
Adding slack variables x_3 and x_4, this problem can be written in canonical form as :

$$\begin{aligned}
\text{Minimise} \quad & -z - x_1 - 5x_2 = 0 \\
\text{Subject to :} \quad & -x_1 + x_2 + x_3 = 1 \\
& 2x_1 + 3x_2 + x_4 = 6 \\
& x_1, x_2, x_3, x_4 \geq 0 \\
& z, x_1, x_2, x_3, x_4 \text{ integer}
\end{aligned}$$

The initial tableau is :

$-z$	x_1	x_2	x_3	x_4	RHS
1	-1	-5	0	0	0
0	-1	1	1	0	1
0	2	3	0	1	6

Using the simplex method the optimal solution of the associated non-integer constrained problem is :

$-z$	x_1	x_2	x_3	x_4	RHS
1	0	0	7/5	6/5	43/5
0	0	1	2/5	1/5	8/5
0	1	0	-3/5	1/5	3/5

This solution is <u>not</u> all integer. Reading the tableau from top to bottom the first row whose right-hand side is not integer is the z-row. Using this, the first cut is :

$$\sum_{j=1}^{n} F(a_{ij})x_j \geq F(b_i)$$

i.e. $\quad 2/5 x_3 + 1/5 x_4 \geq 3/5$

Subtracting a surplus variable x_5 and multiplying through by -1 to make this constraint canonical :

$$-2/5 x_3 - 1/5 x_4 + x_5 = -3/5$$

Adding this constraint to the tableau above :

$-z$	x_1	x_2	x_3	x_4	x_5	RHS
1	0	0	7/5	6/5	0	43/5
0	0	1	2/5	1/5	0	8/5
0	1	0	-3/5	1/5	0	3/5
0	0	0	-2/5	-1/5	1	-3/5

Using the dual simplex method to restore feasibility (using the greyed element in the x_3 column as pivot) :

$-z$	x_1	x_2	x_3	x_4	x_5	RHS
1	0	0	0	1/2	7/2	13/2
0	0	1	0	0	1	1
0	1	0	0	1/2	-3/2	3/2
0	0	0	1	1/2	-5/2	3/2

This solution is still __not__ all integer. Reading the tableau from top to bottom the first row whose right-hand side is not integer is the z-row. Using this, the second cut is :

$$\sum_{j=1}^{n} F(a_{ij}) x_j \geq F(b_i)$$

i.e. $\quad 1/2 x_4 + 1/2 x_5 \geq 1/2$

Subtracting a surplus variable x_6 and multiplying through by -1 to make this constraint canonical :

$$-1/2 x_4 - 1/2 x_5 + x_6 = -1/2$$

Adding this constraint to the tableau above :

$-z$	x_1	x_2	x_3	x_4	x_5	x_6	RHS
1	0	0	0	1/2	7/2	0	13/2
0	0	1	0	0	1	0	1
0	1	0	0	1/2	-3/2	0	3/2
0	0	0	1	1/2	-5/2	0	3/2
0	0	0	0	-1/2	-1/2	1	-1/2

Using the dual simplex method to restore feasibility (using the greyed element in the x_4 column as pivot) :

$-z$	x_1	x_2	x_3	x_4	x_5	x_6	RHS
1	0	0	0	0	3	1	6
0	0	1	0	0	1	0	1
0	1	0	0	0	-2	1	1
0	0	0	1	0	-3	1	1
0	0	0	0	1	1	-2	1

This solution is now all integer. From this tableau the optimal solution of the pure integer linear programming problem is $z = -6$, $x_1 = 1$ and $x_2 = 1$.

Graphical Solution
To illustrate the operation of the cyclic algorithm it will be useful to look at the graphical solution of this problem. To be able to do this the cuts must be expressed in terms of the structural variables x_1 and x_2.

The first cut was :

$$2/5 x_3 + 1/5 x_4 \geq 3/5 \quad i.e. \quad 2x_3 + x_4 \geq 3 \quad \text{---- (1)}$$

From the first canonical form of the problem :

$$x_3 = 1 + x_1 - x_2 \quad \text{and} \quad x_4 = 6 - 2x_1 - 3x_2$$

Substituting these into expression (1) the first cut becomes :

$$2(1 + x_1 - x_2) + (6 - 2x_1 - 3x_2) \geq 3$$

$$\boxed{i.e. \quad x_2 \leq 1}$$

The second cut was :

$$1/2 x_4 + 1/2 x_5 \geq 1/2 \quad i.e. \quad x_4 + x_5 \geq 1 \quad \text{---- (2)}$$

Subtracting a surplus variable x_5 from the first cut :

$$2/5 x_3 + 1/5 x_4 - x_5 = 3/5 \quad i.e. \quad x_5 = 2/5 x_3 + 1/5 x_4 - 3/5$$

Substituting this into expression (2) :

$$x_4 + 2/5 x_3 + 1/5 x_4 - 3/5 \geq 1 \quad i.e. \quad 2x_3 + 6x_4 \geq 8 \quad \text{---- (3)}$$

From the first canonical form of the problem :

138

$$x_3 = 1 + x_1 - x_2 \quad \text{and} \quad x_4 = 6 - 2x_1 - 3x_2$$

Substituting these into expression (3) the second cut becomes :

$$2(1 + x_1 - x_2) + 6(6 - 2x_1 - 3x_2) \geq 8$$

$$\boxed{i.e. \quad x_1 + 2x_2 \leq 3}$$

Using these expressions for the cuts the graphical solution of the pure integer linear programming problem is :

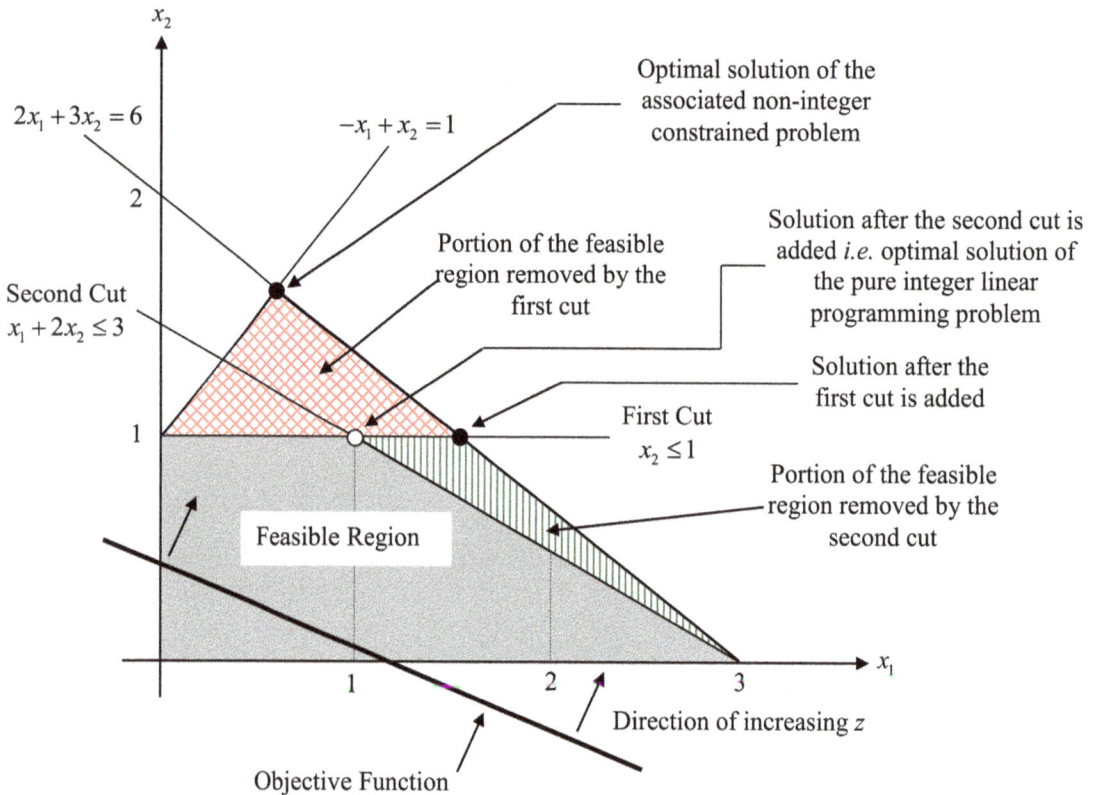

Example 8.3

Use the cyclic algorithm to solve the following pure integer linear programming problem :

$$\text{Minimise} \quad z = 4x_1 + 3x_2$$
$$\text{Subject to :} \quad 2x_1 + 3x_2 \leq 1$$
$$4x_1 + x_2 \geq 1$$
$$x_1, x_2 \geq 0$$
$$z, x_1, x_2 \text{ integer}$$

Solution

This problem can be written in canonical form as :

$$\text{Minimise} \quad -z + 4x_1 + 3x_2 = 0$$
$$-w - 4x_1 - x_2 + x_4 = -1$$
$$\text{Subject to :} \quad 2x_1 + 3x_2 + x_3 = 1$$
$$4x_1 + x_2 - x_4 + a_1 = 1$$
$$x_1, x_2, x_3, x_4, a_1 \geq 0$$
$$z, x_1, x_2, x_3, x_4 \text{ integer}$$

The slack variable x_3 and the surplus variable x_4 are integer constrained. However, the artificial variable a_1 is not integer constrained.

The initial tableau is :

$-z$	$-w$	x_1	x_2	x_3	x_4	a_1	RHS
1	0	4	3	0	0	0	0
0	1	-4	-1	0	1	0	-1
0	0	2	3	1	0	0	1
0	0	4	1	0	-1	1	1

Using the two-phase simplex method the optimal solution of the associated non-integer constrained problem is :

$-z$	x_1	x_2	x_3	x_4	RHS
1	0	2	0	1	-1
0	0	5/2	1	1/2	1/2
0	1	1/4	0	-1/4	1/4

This solution is <u>not</u> all integer. Reading the tableau from top to bottom the first row whose right-hand side is not integer is the first constraint row. Using this, the first cut is :

$$\sum_{j=1}^{n} F(a_{ij})x_j \geq F(b_i)$$

$$\text{i.e.} \quad 1/2\, x_2 + 1/2\, x_4 \geq 1/2$$

Subtracting a surplus variable x_5 and multiplying through by -1 to make this constraint canonical :

$$-1/2\, x_2 - 1/2\, x_4 + x_5 = -1/2$$

Adding this constraint to the tableau above :

$-z$	x_1	x_2	x_3	x_4	x_5	RHS
1	0	2	0	1	0	-1
0	0	5/2	1	1/2	0	1/2
0	1	1/4	0	-1/4	0	1/4
0	0	-1/2	0	-1/2	1	-1/2

Using the dual simplex method to restore feasibility (using the greyed element in the x_4 column as pivot) :

$-z$	x_1	x_2	x_3	x_4	x_5	RHS
1	0	1	0	0	2	-2
0	0	2	1	0	1	0
0	1	1/2	0	0	-1/2	1/2
0	0	1	0	1	-2	1

This solution is still <u>not</u> all integer. Reading the tableau from top to bottom the first row whose right-hand side is not integer is the second constraint row. Using this, the second cut is :

$$\sum_{j=1}^{n} F(a_{ij})x_j \geq F\left(b_i\right)$$

$$i.e. \quad 1/2\,x_2 + 1/2\,x_5 \geq 1/2$$

Subtracting a surplus variable x_6 and multiplying through by -1 to make this constraint canonical :

$$-1/2\,x_2 - 1/2\,x_5 + x_6 = -1/2$$

Adding this constraint to the tableau above :

$-z$	x_1	x_2	x_3	x_4	x_5	x_6	RHS
1	0	1	0	0	2	0	-2
0	0	2	1	0	1	0	0
0	1	1/2	0	0	-1/2	0	1/2
0	0	1	0	1	-2	0	1
0	0	-1/2	0	0	-1/2	1	-1/2

Applying the dual simplex method (using the greyed element in the x_2 column as pivot) gives the tableau :

$-z$	x_1	x_2	x_3	x_4	x_5	x_6	RHS
1	0	0	0	0	1	2	-3
0	0	0	1	0	-1	4	-2
0	1	0	0	0	-1	1	0
0	0	0	0	1	-3	2	0
0	0	1	0	0	1	-2	1

This solution is not feasible. Applying the dual simplex method again (using the greyed element in the x_5 column as pivot) gives the tableau :

$-z$	x_1	x_2	x_3	x_4	x_5	x_6	RHS
1	0	0	1	0	0	6	-5
0	0	0	-1	0	1	-4	2
0	1	0	-1	0	0	-3	2
0	0	0	-3	1	0	-10	6
0	0	1	1	0	0	2	-1

This solution is not feasible *i.e.* the dual simplex method must be applied again. However, in this case the pivot selection rule breaks down indicating that this problem is infeasible. In this context **infeasible** means that the feasible region contains no all-integer point.

General Notes

- When the surplus variable in a cut becomes basic again the cut can be deleted from the tableau. For example, in the tableau above the surplus variable x_5 has become basic again. Hence, had it been possible to select a pivot, the row corresponding to the first cut (*i.e.* row 4) could have been deleted from the tableau.

- The cyclic algorithm can be used for solving pure integer linear programming problems only. However, artificial variables are not integer constrained. Hence, for the method to give an all integer solution, all artificial variables must be non-basic at the end of phase 2. If this is not the case then each basic artificial variable can be made non-basic by choosing the row containing that variable as the pivot row and applying one iteration of the dual simplex method. This operation will maintain optimality and feasibility and will produce a tableau in which the artificial variable is non-basic and can be discarded.

- Integer linear programming problems can arise in which the structural variables are integer constrained but the value of z is not. For example, the structural variables may represent machines and their coefficients in the objective function may be their cost (in £ and pence). To be able to solve a problem of this kind using the cyclic algorithm the objective function must be scaled so that the coefficients are all integer. For example, suppose that the objective function is :

$$z = 2/15 x_1 + 1/6 x_2 + 4/9 x_3$$

Multiplying through by 90 to clear the fractions this expression can be written as :

$$z' = 12x_1 + 15x_2 + 40x_3$$

where $z' = 90z$. If the optimal solution of the problem is $x_1 = 3$, $x_2 = 2$, $x_3 = 1$ and $z' = 106$ then the optimal value of $z = 106/90 = 1.1778$ (to 4 decimal places).

If the objective function cannot be scaled in this way then an alternative method of solution must be used.

8.3 The Branch and Bound Method

One way of solving an integer linear programming problem is to find all integer points within the feasible region, evaluate the objective function at each one and then choose the point that gives the maximum or minimum value of z. This approach is called **enumeration** and can involve a great deal of work since the number of integer points that must be considered grows exponentially with the size of the problem. The **branch and bound method** is a form of enumeration. However, in this method the amount of work required is reduced by ignoring integer points that can be seen to give a "worse" solution than others. The method can be used to solve both pure and mixed integer linear programming problems. However, it is only practical for solving problems in which the integer constrained variables take limited ranges of values otherwise the number of integer points that must be considered remains too high. These ranges are usually determined by financial restrictions, practical considerations, common sense, *etc.*

The first stage when using the branch and bound method is to solve the associated non-integer constrained problem. If all of the integer constrained variables have integer values then the optimal solution has been found and no further work is required. However, if this is not the case then one of the integer constrained variables (that currently has a non-integer value) is selected and a pair of constraints is formed that will force this variable to have an integer value. If the variable selected is x_i then these constraints take the form :

$$x_i \le \text{Int}_{\text{Before}} \quad \text{and} \quad x_i \ge \text{Int}_{\text{After}}$$

where $\text{Int}_{\text{Before}}$ and $\text{Int}_{\text{After}}$ are the integer values on either side of the current non-integer value of the variable. These constraints have the same affect as the **cut** used in the cyclic algorithm *i.e.* they remove the portion of the feasible region that contains the current solution but to ensure that the feasible region still contains the optimal integer solution. The additional constraints are then added separately to the previous optimal tableau. After each constraint is added the problem is put back into canonical form and the dual simplex method is used to restore feasibility. This process is repeated until either the optimal solution is found or until it becomes clear that no optimal integer solution exists.

The progress of the branch and bound method is recorded pictorially on a diagram called a binary tree *i.e.* a tree in which each parent node has two children. The root node of the tree represents the optimal solution of the associated non-integer constrained problem. The other nodes represent the solutions obtained when the additional constraints are added to the problem. The value of z at each node is called an **upper bound** on the optimal solution (in a maximisation problem) or a **lower bound** on the optimal solution (in a minimisation problem). To reduce the amount of work required to find the optimal solution the tree is expanded only as far as necessary. Whenever there is a choice for the next node to expand the most "promising" node only is selected. This is the unexpanded node that has the largest upper bound or smallest lower bound. If a node is not going to be expanded further then it is usual to draw a line below it on the tree to indicate that the path is closed.

The Procedure

To solve an integer linear programming problem using the branch and bound method :

1. Solve the associated non-integer constrained problem. If all of the integer constrained variables have integer values then the optimal solution has been found and no further work is required. Otherwise :

2. Draw the root node of the binary tree and annotate it with the optimal solution of the associated non-integer constrained problem.

3. Select one of the integer constrained variables that currently has a non-integer value and form the pair of constraints that will force this variable to have an integer value.

4. Add nodes to the binary tree (with branches connecting them to the parent node) to represent the solutions obtained when the additional constraints are added to the problem.

5. Incorporate the type 1 constraint *i.e.* convert the constraint into an equation by adding a slack variable. Add this constraint to the tableau corresponding to the parent node. Put the problem back into canonical form by subtracting the structural constraint containing the variable selected at step 3. from the new constraint. Use the dual simplex method to restore feasibility. Annotate the corresponding node in the binary tree with the solution obtained. If the solution is integer or if the problem is infeasible, draw a line below the corresponding node to indicate that the path to that node is closed.

6. Incorporate the type 3 constraint *i.e.* convert the constraint into an equation by subtracting a surplus variable. Multiply through the constraint by -1. Add this constraint to the tableau corresponding to the parent node. Put the problem back into canonical form by adding the structural constraint containing the variable selected at step 3. to the new constraint. Use the dual simplex method to restore feasibility. Annotate the corresponding node in the binary tree with the solution obtained. If the solution is integer or if the problem is infeasible, draw a line below the corresponding node to indicate that the path to that node is closed.

7. Examine the binary tree . If the tree is complete *i.e.* if no other nodes need to be explored then stop and extract the optimal solution. This will be given at the node that has the largest upper bound (in a maximisation problem) or the smallest lower bound (in a minimisation problem). Otherwise :

8. Choose the most "promising" node to expand and go back to step 3.

Example 8.4
An airline, LRF Airways wishes to purchase a number of aircraft to operate over the London to New York route. There are two suitable types of aircraft available only, the Boeing 797 and the Airbus A400. The Boeing carries 180 passengers, requires a crew of 12 and uses 1500 gallons of fuel each trip. The Airbus carries 200 passengers, requires a crew of 20 and uses 1400 gallons of fuel each trip. The airline has 70 crew members available but not all of them need to be employed. It forecasts that there will be 900 passengers each trip and wishes to minimise the total amount of fuel used on each trip.

1) Formulate an integer linear programming model from which the airline can determine the number of aircraft of each type to purchase.

2) Explain why the branch and bound method is a practical method for solving the integer linear programming model from 1).

3) Solve the integer linear programming model from 1) using the branch and bound method.

Solution
1) Let :

z be the total amount of fuel used on each trip.

x_1 be the number of Boeing aircraft purchased.

x_2 be the number of Airbus aircraft purchased.

Then, from the information given :

The total amount of fuel used on each trip is :

$$z = 1500x_1 + 1400x_2$$

The passenger constraint is :

$$180x_1 + 200x_2 \geq 900$$

The aircrew constraint is :

$$12x_1 + 20x_2 \leq 70$$

Combining these expressions the required integer linear programming model is :

Minimise $\quad z = 1500x_1 + 1400x_2$
Subject to : $\quad 180x_1 + 200x_2 \geq 900 \quad$ ---- (1)
$\quad\quad\quad\quad 12x_1 + 20x_2 \leq 70 \quad$ ---- (2)
$\quad\quad\quad\quad x_1, x_2 \geq 0$
$\quad\quad\quad\quad x_1, x_2$ integer

145

2) Consider constraint (2). By setting $x_1 = 0$ and then $x_2 = 0$ it can be seen that the number of Boeing aircraft must be less than 6 and the number of Airbus aircraft must be less than 4. Hence, x_1 must take one of the integer values 0, 1, 2, 3, 4, 5 and x_2 must take one of the integer values 0, 1, 2, 3. Since the integer constrained variables take limited ranges of values the branch and bound method is a practical method for solving this problem.

3) Using a surplus variable x_3, a slack variable x_4 and an artificial variable a_1 the associated non-integer constrained problem can be written in canonical form as :

$$\text{Minimise} \qquad -z + 1500x_1 + 1400x_2 = 0$$
$$-w - 180x_1 - 200x_2 + x_3 = -900$$
$$\text{Subject to :} \qquad 180x_1 + 200x_2 - x_3 + a_1 = 900$$
$$12x_1 + 20x_2 + x_4 = 70$$
$$x_1, x_2, x_3, x_4, a_1 \geq 0$$

The initial tableau is :

$-z$	$-w$	x_1	x_2	x_3	x_4	a_1	RHS
1	0	1500	1400	0	0	0	0
0	1	-180	-200	1	0	0	-900
0	0	180	200	-1	0	1	900
0	0	12	20	0	1	0	70

Using the two-phase simplex method the optimal solution of this problem is :

$-z$	x_1	x_2	x_3	x_4	RHS
1	0	0	11	40	-7100
0	1	0	-1/60	-1/6	10/3
0	0	1	1/100	3/20	3/2

i.e. $z = 7100$, $x_1 = 10/3$ and $x_2 = 3/2$. Note : The non-basic artificial variable a_1 and the w-row have been removed from this tableau. The root node of the binary tree is :

LB = 7100
$x_1 = 10/3$
$x_2 = 3/2$

Root Node

This solution is <u>not</u> all integer. The integer constrained variable x_1 currently has a non-integer value. The additional constraints required to force this variable to have an integer value are :

$$x_1 \leq 3 \quad \text{and} \quad x_1 \geq 4$$

146

Adding appropriate branches to the binary tree :

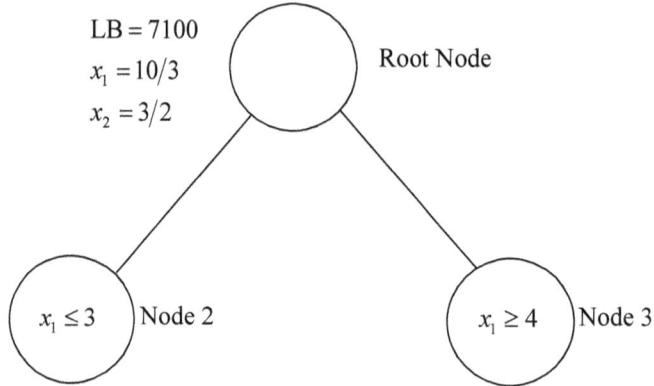

Node 2

Adding a slack variable x_5, the constraint $x_1 \leq 3$ becomes $x_1 + x_5 = 3$. Adding this constraint to the tableau corresponding to the parent node (*i.e.* the root node) :

$-z$	x_1	x_2	x_3	x_4	x_5	**RHS**
1	0	0	11	40	0	-7100
0	1	0	$-1/60$	$-1/6$	0	10/3
0	0	1	$1/100$	$3/20$	0	3/2
0	1	0	0	0	1	3

Subtracting row 2 (*i.e.* the structural constraint containing x_1) from row 4 (*i.e.* the new constraint), the problem can be rewritten in canonical form as :

$-z$	x_1	x_2	x_3	x_4	x_5	**RHS**
1	0	0	11	40	0	-7100
0	1	0	$-1/60$	$-1/6$	0	10/3
0	0	1	$1/100$	$3/20$	0	3/2
0	0	0	$1/60$	$1/6$	1	$-1/3$

The dual simplex method must now be used to restore feasibility. However, in this case the pivot selection rule breaks down indicating that this problem is infeasible. Hence, node 2 can be closed. Updating the binary tree :

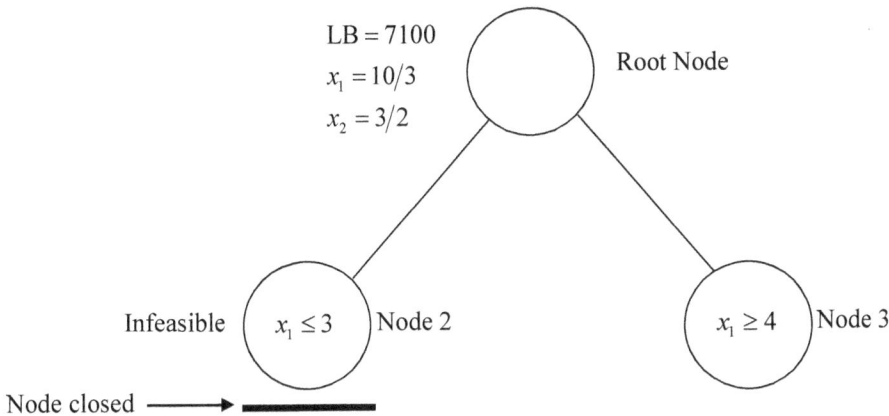

Node 3

Subtracting a surplus variable x_6 and multiplying through by -1 the constraint $x_1 \geq 4$ becomes $-x_1 + x_6 = -4$. Adding this constraint to the tableau corresponding to the parent node (*i.e.* the root node) :

$-z$	x_1	x_2	x_3	x_4	x_6	**RHS**
1	0	0	11	40	0	-7100
0	1	0	$-1/60$	$-1/6$	0	$10/3$
0	0	1	$1/100$	$3/20$	0	$3/2$
0	-1	0	0	0	1	-4

Adding row 2 (*i.e.* the structural constraint containing x_1) to row 4 (*i.e.* the new constraint), the problem can be rewritten in canonical form as :

$-z$	x_1	x_2	x_3	x_4	x_6	**RHS**
1	0	0	11	40	0	-7100
0	1	0	$-1/60$	$-1/6$	0	$10/3$
0	0	1	$1/100$	$3/20$	0	$3/2$
0	0	0	$-1/60$	$-1/6$	1	$-2/3$

Using the dual simplex method to restore feasibility (using the greyed element in the x_4 column as pivot) gives the solution :

$-z$	x_1	x_2	x_3	x_4	x_6	**RHS**
1	0	0	7	0	240	-7260
0	1	0	0	0	-1	4
0	0	1	$-1/200$	0	$9/10$	$9/10$
0	0	0	$1/10$	1	-6	4

i.e. $z = 7260$, $x_1 = 4$ and $x_2 = 9/10$. Updating the binary tree :

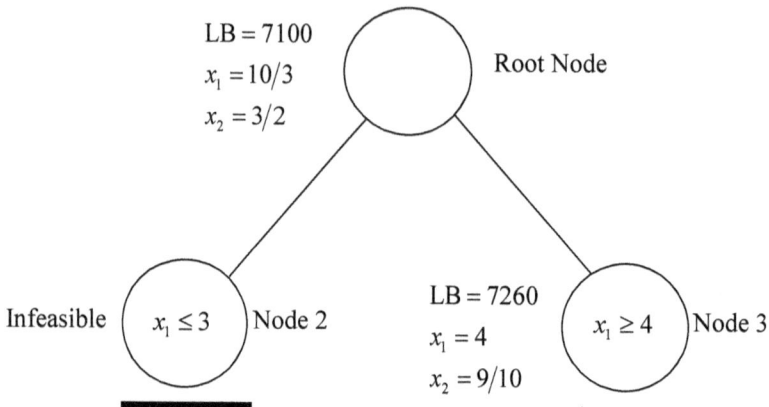

This solution is <u>not</u> all integer. The integer constrained variable x_2 currently has a non-integer value. The additional constraints required to force this variable to have an integer value are :

$$x_2 \leq 0 \quad \text{and} \quad x_2 \geq 1$$

Adding appropriate branches to the binary tree :

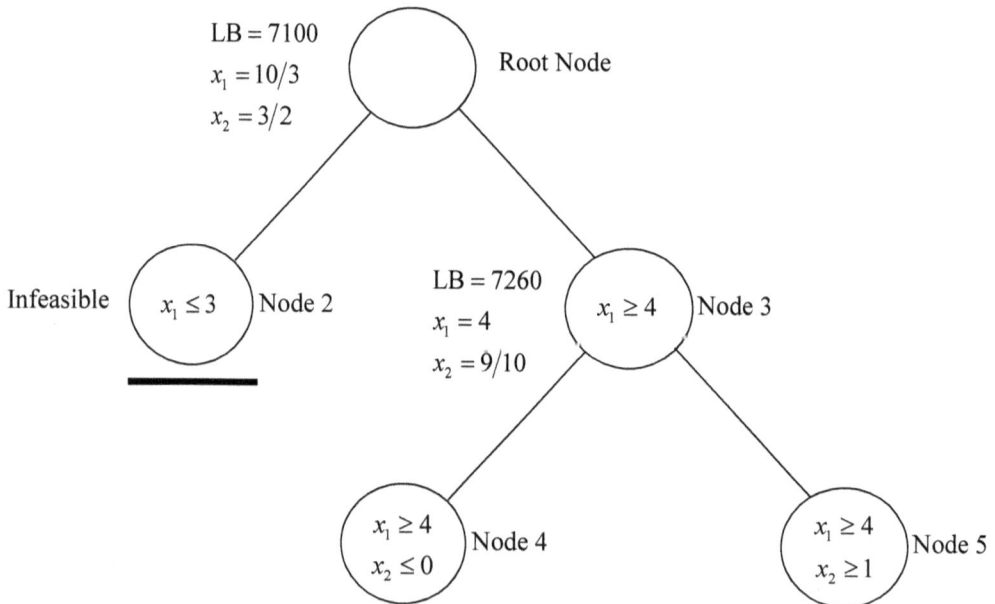

Node 4

Since x_2 is constrained to be non-negative the only possible solution is $x_2 = 0$. In this case the original integer linear programming model reduces to :

$$\text{Minimise} \qquad z = 1500x_1$$

$$\text{Subject to :} \qquad 180x_1 \geq 900 \quad \text{---- (3)}$$

$$12x_1 \leq 70 \quad \text{---- (4)}$$

$$x_1 \geq 0$$

$$x_1 \text{ integer}$$

Constraint (3) $\Rightarrow x_1 \geq 5$ and constraint (4) $\Rightarrow x_1 \leq 5$. Hence, the only possible solution of this problem is $x_1 = 5$. This solution also satisfies the other constraint at this node *i.e.* $x_1 \geq 4$. Substituting $x_1 = 5$ and $x_2 = 0$ into the objective function, $z = 7500$. This solution is all integer. Hence, node 4 can be closed. Updating the binary tree :

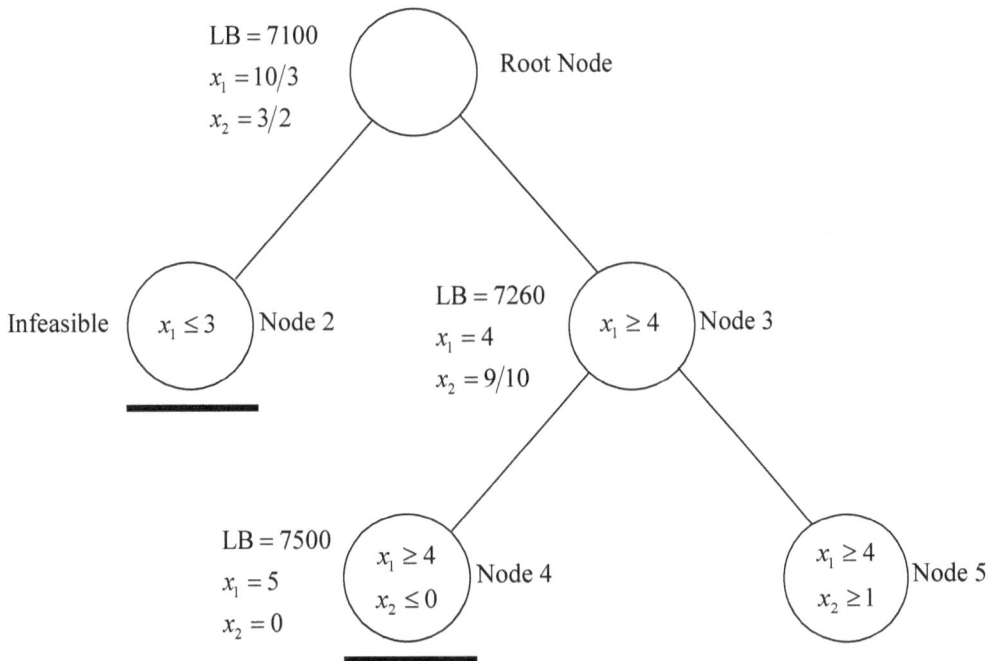

Node 5

Subtracting a surplus variable x_7 and multiplying through by -1 the constraint $x_2 \geq 1$ becomes $-x_2 + x_7 = -1$. Adding this constraint to the tableau corresponding to the parent node (*i.e.* node 3) :

$-z$	x_1	x_2	x_3	x_4	x_6	x_7	RHS
1	0	0	7	0	240	0	-7260
0	1	0	0	0	-1	0	4
0	0	1	$-1/200$	0	$9/10$	0	$9/10$
0	0	0	$1/10$	1	-6	0	4
0	0	-1	0	0	0	1	-1

Adding row 3 (*i.e.* the structural constraint containing x_2) to row 5 (*i.e.* the new constraint), the problem can be rewritten in canonical form as :

$-z$	x_1	x_2	x_3	x_4	x_6	x_7	RHS
1	0	0	7	0	240	0	-7260
0	1	0	0	0	-1	0	4
0	0	1	$-1/200$	0	9/10	0	9/10
0	0	0	1/10	1	-6	0	4
0	0	0	$-1/200$	0	9/10	1	$-1/10$

Using the dual simplex method to restore feasibility (using the greyed element in the x_3 column as pivot) gives the solution :

$-z$	x_1	x_2	x_3	x_4	x_6	x_7	RHS
1	0	0	0	0	1500	1400	-7400
0	1	0	0	0	-1	0	4
0	0	1	0	0	0	-1	1
0	0	0	0	1	12	20	2
0	0	0	1	0	-180	-200	20

i.e. $z = 7400$, $x_1 = 4$ and $x_2 = 1$. This solution <u>is</u> all integer. Hence, node 5 can be closed. Updating the binary tree :

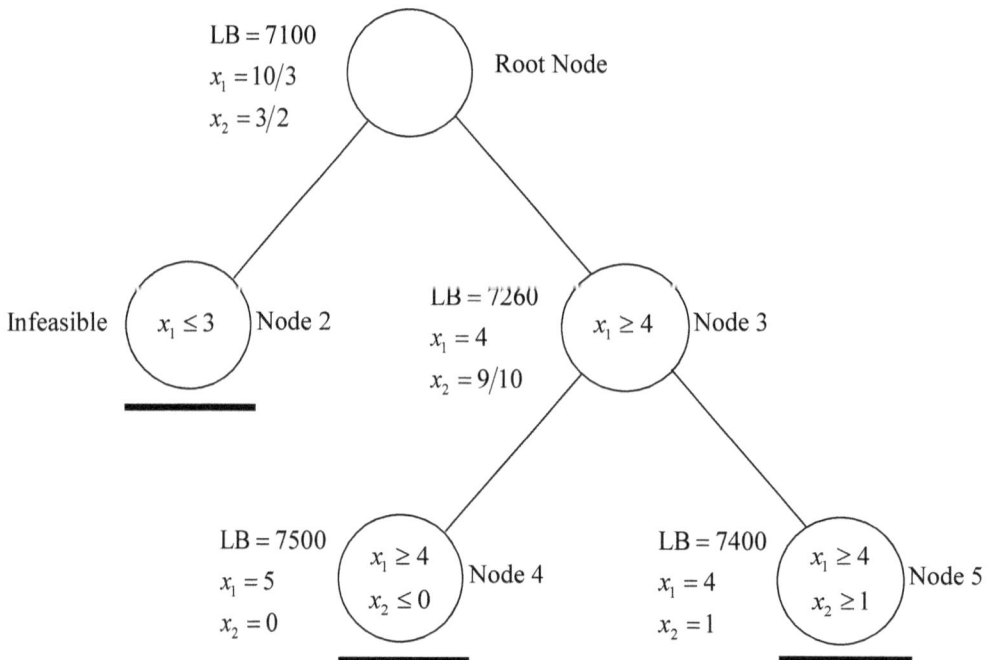

Since the binary tree is complete *i.e.* since all nodes are closed, no further work is required. The node with the smallest lower bound is node 5. The solution here is $z = 7400$, $x_1 = 4$ and $x_2 = 1$.

Hence, LRF Airways should purchase 4 Boeing aircraft and 1 Airbus aircraft. The total amount of fuel used on each trip will then be 7400 gallons.

Graphical Solution

To illustrate the operation of the branch and bound method it will be useful to look at the graphical solution of this problem.

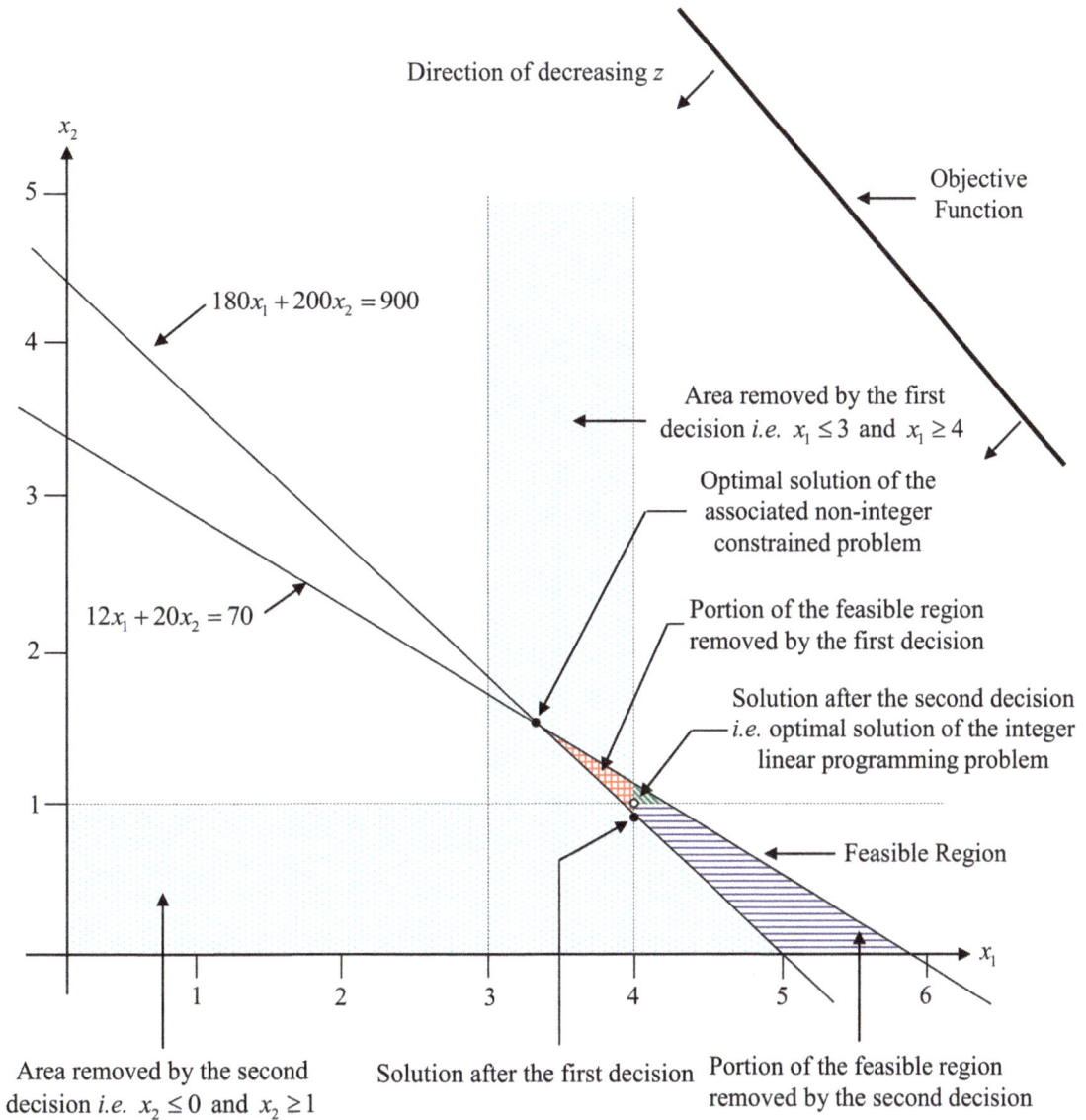

Direction of decreasing z

Objective Function

x_2

5 —

$180x_1 + 200x_2 = 900$

4 —

Area removed by the first decision *i.e.* $x_1 \le 3$ and $x_1 \ge 4$

Optimal solution of the associated non-integer constrained problem

3 —

$12x_1 + 20x_2 = 70$

Portion of the feasible region removed by the first decision

2 —

Solution after the second decision *i.e.* optimal solution of the integer linear programming problem

1 —

Feasible Region

1   2   3   4   5   6 x_1

Area removed by the second decision *i.e.* $x_2 \le 0$ and $x_2 \ge 1$

Solution after the first decision

Portion of the feasible region removed by the second decision

8.4 Suboptimisation

The amount of work required by the branch and bound method can be reduced further by using a technique called suboptimisation. Here, the method is terminated when the "best" integer solution found so far is "nearly optimal" *i.e.* when the upper/lower bound in this solution is sufficiently close to (*e.g.* within 0.5% of) the upper/lower bound at the most "promising" unexplored node. To illustrate this procedure consider the following fictitious binary tree for a minimisation problem :

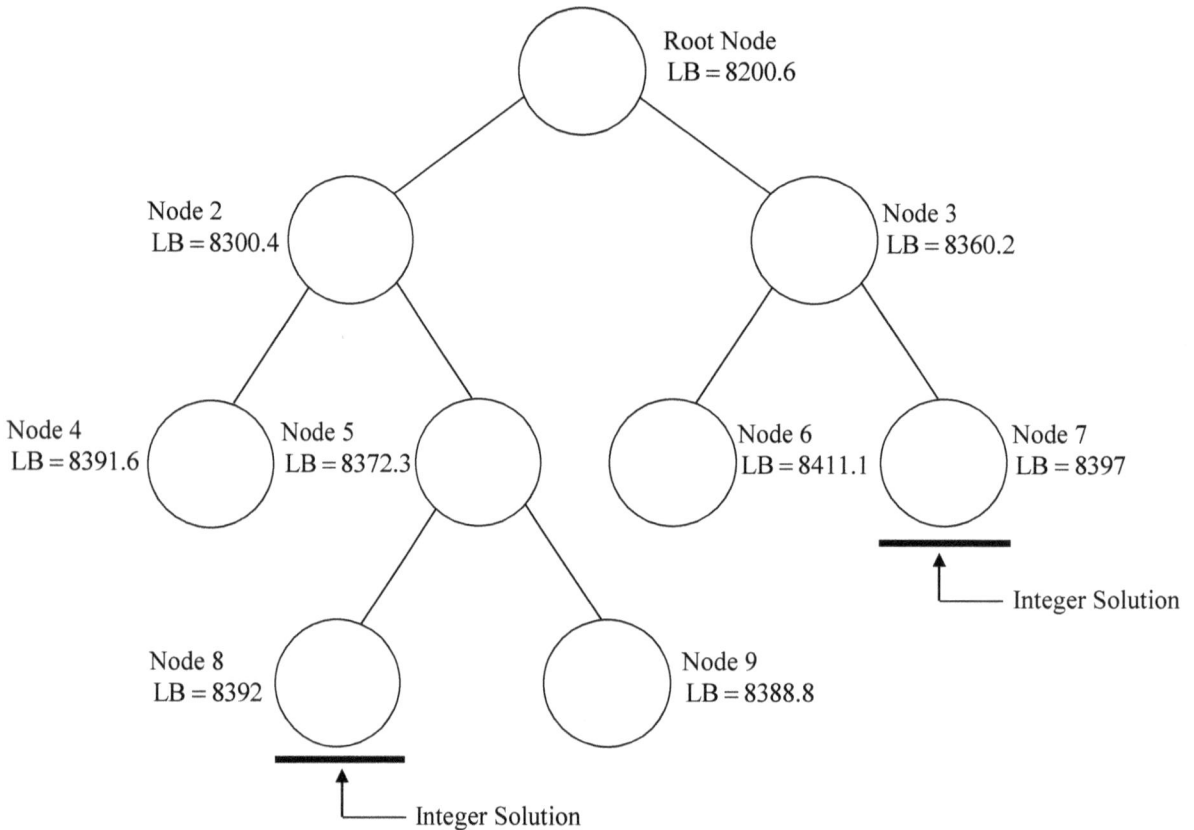

To find the optimal solution of this problem nodes 4, 6 and 9 need to be explored further. However :

- The "best" integer solution found so far is node 8. This has a lower bound of 8392.

- The most "promising" unexplored node is node 9. This has a lower bound of 8388.3.

- The percentage difference between these lower bounds is :

$$\frac{8392 - 8388.3}{8392} \times \frac{100}{1} \cong 0.0441\%$$

Hence, using suboptimisation the method can terminate at this point. The optimal solution is taken to be the integer solution at node 8.

153

8.5 Exercises 8

1. Use the cyclic algorithm to solve the following pure integer linear programming problem :

$$\begin{aligned} \text{Maximise} \quad & z = x_1 + 3x_2 \\ \text{Subject to :} \quad & x_1 + x_2 \le 10 \\ & 3x_1 + 11x_2 \le 33 \\ & x_1, x_2 \ge 0 \\ & z, x_1, x_2 \text{ integer} \end{aligned}$$

2. Use the cyclic algorithm to solve the following pure integer linear programming problem :

$$\begin{aligned} \text{Maximise} \quad & z = 350x_1 + 300x_2 \\ \text{Subject to :} \quad & x_1 + x_2 \le 200 \\ & 9x_1 + 6x_2 \le 1566 \\ & 12x_1 + 16x_2 \le 2880 \\ & x_1, x_2 \ge 0 \\ & z, x_1, x_2 \text{ integer} \end{aligned}$$

3. The Kendal Post is a free local newspaper that is issued weekly. The printers receive large 48-inch wide rolls of paper from their supplier that must be cut into <u>two</u> sizes to produce each edition. For each issue the printers require twenty 25-inch rolls for the main pages of the newspaper and fifty 21-inch rolls for the inserts and advertisements. The cutting process produces waste and naturally the management at the newspaper wish to minimise this waste in order to keep their production costs to a minimum.

   (i) Formulate an integer linear programming model from which the management at the newspaper can determine the optimal solution of this paper trimming problem. Hint : Identify the possible cutting patterns and the corresponding trim wastes. It will be useful to summarise this information in a table before formulating the integer linear programming model. See question 6. in these exercises for an example of a table of this kind.

   (ii) Solve the integer linear programming model from 3(i) using the cyclic algorithm.

4. Use the branch and bound method to solve the following pure integer linear programming problem :

$$\begin{aligned} \text{Minimise} \quad & z = x_1 + 3x_2 \\ \text{Subject to :} \quad & x_1 + x_2 \ge 5 \\ & 1/5\,x_1 + x_2 \ge 3 \\ & 1/3\,x_1 + x_2 \le 5 \\ & x_1, x_2 \ge 0 \\ & z, x_1, x_2 \text{ integer} \end{aligned}$$

5. Use the branch and bound method to solve the following pure integer linear programming problem :

$$\text{Maximise} \quad z = 2x_1 + 3x_2$$

$$\text{Subject to :} \quad x_1 + 3x_2 \leq 8.25$$

$$2.5x_1 + x_2 \leq 8.75$$

$$x_1, x_2 \geq 0$$

$$z, x_1, x_2 \text{ integer}$$

6. The Daily Crucible is a national newspaper with a daily circulation of over 100,000 copies. For this paper the printers receive large $215cm$ wide rolls of paper from their supplier that must be cut into <u>three</u> sizes for various parts of the paper. Each day the printers require one hundred and eighty $64cm$ rolls, ninety $60cm$ rolls and ninety $35cm$ rolls. As before the cutting process produces waste that the management at the newspaper wish to minimise. The possible cutting patterns and trim wastes in this case are :

Cutting Pattern	1	2	3	4	5	6	7	8	9	10
Number of **64cm** Rolls	3	2	2	1	1	1	0	0	0	0
Number of **60cm** Rolls	0	1	0	2	1	0	3	2	1	0
Number of **35cm** Rolls	0	0	2	0	2	4	1	2	4	6
Trim Waste (*cm*)	23	27	17	31	21	11	0	25	15	5

(i) Formulate an integer linear programming model from which the management at the newspaper can determine the optimal solution of this paper trimming problem.

(ii) Solve the integer linear programming model from 6(i) using the Solver tool in *Excel*. An example of using this tool is given in Chapter 10. Is the solution obtained sensible ? If not, state how a better solution can be obtained.

9. Post-Optimal Analysis

9.1 Introduction
Post-optimal analysis is concerned with investigating the effects of changes in the parameters of a linear programming problem. One way of determining these effects is to resolve the linear programming problem using the new data. However, this method usually involves more work than the procedures described in this chapter.

Preliminary Notes
- Consider the following general linear programming problem :

$$
\begin{aligned}
\text{Minimise} \quad & z = c_0 + \underline{c}\,\underline{x} \\
\text{Subject to :} \quad & A\underline{x} \leq \underline{b} \\
& \underline{x} \geq \underline{0}
\end{aligned}
$$

where $\underline{b} \geq \underline{0}$, $\underline{c}$ is a row vector, A is a matrix, $\underline{x}$ is a column vector and $\underline{b}$ is a column vector. In Chapter 4 it was shown that the optimal simplex tableau for this problem can be written in matrix form as :

$$
\begin{bmatrix} 1 & \underline{0} & \underline{c}_N - \underline{c}_B B^{-1} N \\ \underline{0} & I & B^{-1} N \end{bmatrix}
\begin{bmatrix} -z \\ \underline{x}_B \\ \underline{x}_N \end{bmatrix}
=
\begin{bmatrix} -c_0 - \underline{c}_B B^{-1} \underline{b} \\ B^{-1} \underline{b} \end{bmatrix}
$$

where $\underline{x}_B$ contains the basic variables, $\underline{x}_N$ contains the non-basic variables, $\underline{c}_B$ contains the coefficients of the basic variables in the objective function, $\underline{c}_N$ contains the coefficients of the non-basic variables in the objective function, B contains the columns of the matrix A associated with the basic variables (in the order in which they occur in the matrix A) and N contains the columns of the matrix A associated with the non-basic variables (in the order in which they occur in the matrix A).

The current solution of this problem is optimal if the coefficients of the non-basic variables in the objective function row are non-negative. It can be seen from the expression above that this condition is satisfied if $(\underline{c}_N - \underline{c}_B B^{-1} N) \geq \underline{0}$.

- In the procedures described in this chapter :

  $\underline{c}_B^*$ contains the z-row coefficients from the <u>first canonical form</u> of the problem that correspond to the **basic** variables in the optimal tableau.

  $\underline{c}_N^*$ contains the z-row coefficients from the <u>first canonical form</u> of the problem that correspond to the **non-basic** variables in the optimal tableau.

  B^* contains the columns of the matrix A from the <u>first canonical form</u> of the problem that correspond to the **basic** variables in the optimal tableau.

N^* contains the columns of the matrix A from the <u>first canonical form</u> of the problem that correspond to the **non-basic** variables in the optimal tableau.

- The theorems used in this chapter are the the duality theorems given in Chapter 7.

- When performing a post-optimal analysis it is often necessary to find the matrix B^{*-1}. One way of doing this is to write down the matrix B^* (using the definition given above) and then to calculate its inverse using one of the standard procedures. However, it is more efficient to find B^{*-1} using one of the following methods :

Case 1 : Using the Simplex Methods
If the linear programming problem has been solved using one of the simplex methods the matrix B^{*-1} can be extracted from the optimal tableau. The columns of B^{*-1} are the columns of the optimal tableau that correspond to the basic variables in the <u>initial tableau</u> ordered so that the columns associated with these variables in the <u>initial tableau</u> form an identity matrix.

Note
In phase 2 of the two-phase simplex method $(-w)$ is treated as a non-negative variable. Hence, $(-w)$ can be treated as a basic variable in the first canonical form of the problem.

Example 9.1
Consider the following linear programming problem :

$$\begin{aligned}
\text{Minimise} \quad & z = 2x_1 - x_2 \\
\text{Subject to :} \quad & x_2 \leq 1 \\
& x_1 \leq 1 \\
& x_1 - x_2 = 1/2 \\
& x_1 + x_2 \geq 1 \\
& x_1, x_2 \geq 0
\end{aligned}$$

This problem can be written in canonical form as :

$$\begin{aligned}
\text{Minimise} \quad & -z + 2x_1 - x_2 = 0 \\
& -w - 2x_1 + x_5 = 3/2 \\
\text{Subject to :} \quad & x_2 + x_3 = 1 \\
& x_1 + x_4 = 1 \\
& x_1 - x_2 + a_1 = 1/2 \\
& x_1 + x_2 - x_5 + a_2 = 1 \\
& x_1, x_2, x_3, x_4, x_5, a_1, a_2 \geq 0
\end{aligned}$$

The initial two-phase simplex tableau is :

(1)	(2)	(3)	(4)	(5)	(6)	(7)	(8)	(9)	(10)
$-z$	$-w$	x_1	x_2	x_3	x_4	x_5	a_1	a_2	**RHS**
1	0	2	-1	0	0	0	0	0	0
0	1	-2	0	0	0	1	0	0	$-3/2$
0	0	0	1	1	0	0	0	0	1
0	0	1	0	0	1	0	0	0	1
0	0	1	-1	0	0	0	1	0	1/2
0	0	1	1	0	0	-1	0	1	1

The columns associated with the basic variables are (2), (5), (6), (8) and (9). The order in which these columns form an identity matrix is also (2), (5), (6), (8) and (9).

The optimal two-phase simplex tableau is :

(1)	(2)	(3)	(4)	(5)	(6)	(7)	(8)	(9)	(10)
$-z$	$-w$	x_1	x_2	x_3	x_4	x_5	a_1	a_2	**RHS**
1	0	0	0	0	0	1/2	$-3/2$	$-1/2$	$-5/4$
0	1	0	0	0	0	0	1	1	0
0	0	0	0	1	0	1/2	1/2	$-1/2$	3/4
0	0	0	0	0	1	1/2	$-1/2$	$-1/2$	1/4
0	0	1	0	0	0	$-1/2$	1/2	1/2	3/4
0	0	0	1	0	0	$-1/2$	$-1/2$	1/2	1/4

Extracting the variable coefficients from columns (2), (5), (6), (8) and (9), the matrix B^{*-1} is :

$$B^{*-1} = \begin{bmatrix} 1 & 0 & 0 & 1 & 1 \\ 0 & 1 & 0 & 1/2 & -1/2 \\ 0 & 0 & 1 & -1/2 & -1/2 \\ 0 & 0 & 0 & 1/2 & 1/2 \\ 0 & 0 & 0 & -1/2 & 1/2 \end{bmatrix}$$

Note
To be able to find the matrix B^{*-1} in this way the artificial variables must **not** be discarded as soon as they become non-basic.

Case 2 : Using the Revised Simplex Methods
If the linear programming problem has been solved using one of the revised simplex methods the matrix B^{*-1} can be extracted from the optimal transformation matrix. This matrix has the general form :

$$T = \begin{bmatrix} 1 & -\underline{c}_B^* B^{*-1} \\ \underline{0} & B^{*-1} \end{bmatrix}$$

$$\longleftarrow B^{*-1}$$

Example 9.2

Consider the following linear programming problem :

$$\begin{aligned}
\text{Minimise} \quad & z = -5x_1 - 2x_2 - 2x_3 \\
\text{Subject to :} \quad & x_1 + x_2 + x_3 = 1 \\
& x_1 + 2x_2 - x_3 = 2 \\
& x_1, x_2, x_3 \geq 0
\end{aligned}$$

This problem can be written in canonical form as :

$$\begin{aligned}
\text{Minimise} \quad & -z - 5x_1 - 2x_2 - 2x_3 = 0 \\
& -w - 2x_1 - 3x_2 = -3 \\
\text{Subject to :} \quad & x_1 + x_2 + x_3 + a_1 = 1 \\
& x_1 + 2x_2 - x_3 + a_2 = 2 \\
& x_1, x_2, x_3, a_1, a_2 \geq 0
\end{aligned}$$

The optimal transformation matrix for this problem is :

$$T_2 = \begin{bmatrix} 1 & 3 & 11 & 0 \\ 0 & 1 & 3 & 0 \\ 0 & -1 & -2 & 0 \\ 0 & 1 & 1 & 1 \end{bmatrix}$$

Hence, the matrix B^{*-1} is :

$$B^{*-1} = \begin{bmatrix} 1 & 3 & 0 \\ -1 & -2 & 0 \\ 1 & 1 & 1 \end{bmatrix}$$

9.2 Changes in the Vector $\underline{b}$

Changing the b coefficient on the right-hand side of a constraint moves that constraint **parallel** to itself in the feasible region.

Case 1 : Changes that <u>do not</u> affect the optimal basis

The permissible change in b_i *i.e.* the maximum change that will **<u>not</u>** affect the optimal basis depends upon whether the corresponding constraint is **slack** or **tight** at the optimum.

Consider the Toy Manufacturing Problem from Chapter1 *i.e.*

$$\begin{aligned}
\text{Maximise :} \quad & z = 8x_1 + 5x_2 \\
\text{Subject to :} \quad & 3x_1 + x_2 \leq 12 \\
& x_1 \leq 3 \\
& x_1 + x_2 \leq 7 \\
& x_1, x_2 \geq 0
\end{aligned}$$

Slack Constraints

Slack constraints do **not** pass through the optimal vertex. For a constraint of this type the value of b_i can be varied without affecting the optimal basis provided that the change does not cause the constraint pass beyond the optimal vertex. For example, consider the graphical solution of the Toy Manufacturing problem *i.e.*

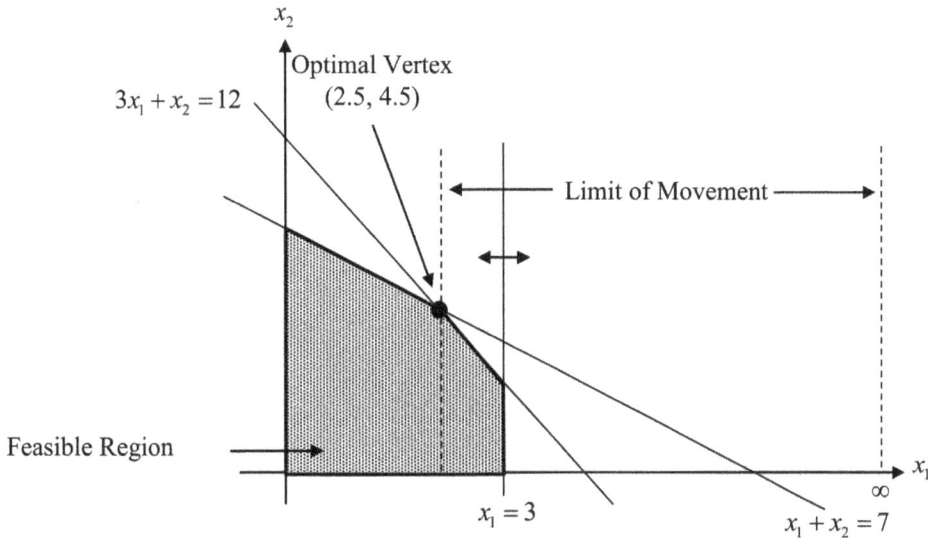

The type 1 constraint $x_1 \leq 3$ is slack at the optimum. From the diagram above it can be seen that the requirement b_2 can be varied without causing the constraint to pass beyond the optimal vertex *i.e.* without affecting the optimal basis provided that the constraint remains within the limits shown.

In general, the rules for varying the right-hand sides of **slack** constraints are as follows :

Type 1 Constraints
If a **type 1 constraint** is slack at the optimum, the b-coefficient on the right-hand side of that constraint can be increased by ∞ and reduced by x_s (where x_s is the optimal value of the **slack** variable in that constraint) without affecting the optimal basis.

Type 3 Constraints
If a **type 3 constraint** is slack at the optimum, the b-coefficient on the right-hand side of that constraint can be can be reduced by ∞ and increased by x_s (where x_s is the optimal value of the **surplus** variable in that constraint) without affecting the optimal basis.

Tight Constraints
Tight constraints **do** pass through the optimal vertex. For a constraint of this type the value of b_i can be varied without affecting the optimal basis provided that the change does not cause a different set of constraints to intersect at the optimal vertex. For example, consider the graphical solution of the Toy Manufacturing problem *i.e.*

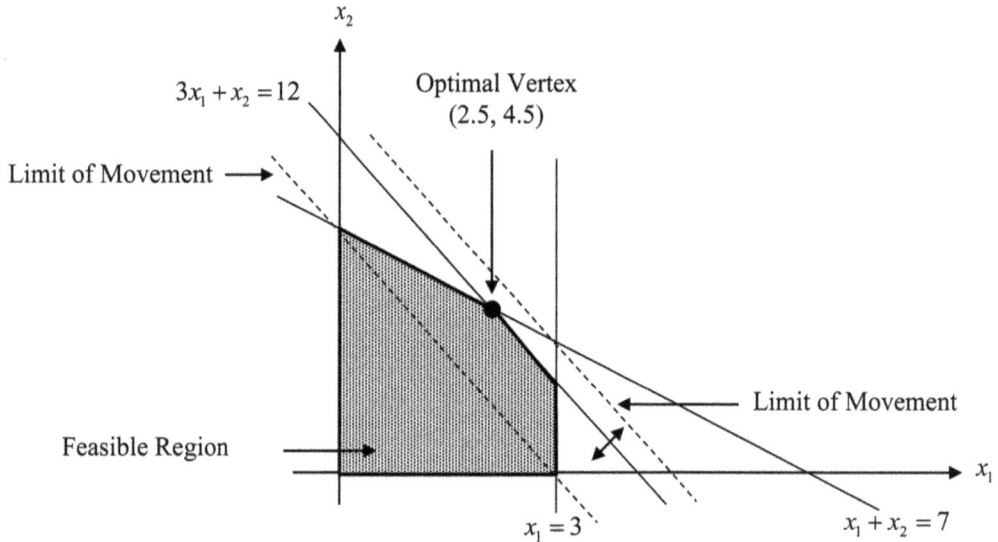

The optimal vertex occurs at the intersection of the constraints $3x_1 + x_2 \leq 12$ and $x_1 + x_2 \leq 7$.

The type 1 constraint $3x_1 + x_2 \leq 12$ is tight at the optimum. From the diagram above it can be seen that the requirement b_1 can be varied without causing a different set of constraints to intersect at the optimal vertex *i.e.* without affecting the optimal basis provided that the constraint remains within the limits shown.

In general, the rules for varying the right-hand sides of **tight** constraints are as follows :

Type 1 Constraints
If a **type 1 constraint** is tight at the optimum, the b-coefficient on the right-hand side of that constraint can be increased by :

$$\min\left\{\frac{b_i}{-a_{is}} \mid a_{is} < 0; 1 \leq l \leq m\right\}$$

and reduced by :

$$\min\left\{\frac{b_i}{a_{is}} \mid a_{is} > 0; 1 \leq i \leq m\right\}$$

(where x_s is the **slack variable** in that constraint) without affecting the optimal basis.

Type 3 Constraints
If a **type 3 constraint** is tight at the optimum, the b-coefficient on the right-hand side of that constraint can be increased by :

$$\min\left\{\frac{b_i}{a_{is}} \mid a_{is} > 0; 1 \leq i \leq m\right\}$$

161

and reduced by :

$$\min\left\{\frac{b_i}{-a_{is}}\;\middle|\;a_{is}<0;1\le i\le m\right\}$$

(where x_s is the **surplus variable** in that constraint) without affecting the optimal basis.

Notes
- **Type 2 constraints** do not have slack or surplus variables associated with them. Hence, the b-coefficient on the right-hand side of a type 2 constraint cannot be varied without affecting the optimal basis.

- Although the b-coefficients can be varied within the ranges given above without affecting the optimal basis of the primal, large increases in the b-coefficients within these ranges can affect the optimal basis of the dual.

The New Value of z
Suppose that a requirement b_i changes by a small amount δb_i. Then, by the total differential the new optimal value of z i.e. z^+ is given by :

$$z^+ = z^* + \frac{\partial z^*}{\partial b_i}\delta b_i$$

where z^* is the optimal value of z before the change is made. By Theorem 6 the value of the partial derivative is the optimal value of the ith dual variable i.e. y_i.

Example 9.3
Consider the following linear programming problem :

$$\begin{aligned}
\text{Minimise} \quad & z = -x_1 - 2x_2 \\
\text{Subject to :} \quad & x_1 + x_2 \le 10 \quad \text{---- (1)} \\
& -x_1 + 4x_2 \le 16 \quad \text{---- (2)} \\
& 2x_1 + 5x_2 \le 30 \quad \text{---- (3)} \\
& x_1 - x_2 \le 6 \quad \text{---- (4)} \\
& x_1, x_2 \ge 0
\end{aligned}$$

Adding slack variables x_3, x_4, x_5 and x_6 to constraints (1), (2), (3) and (4) respectively this problem can be written in canonical form as :

$$\begin{aligned}
\text{Minimise} \quad & -z - x_1 - 2x_2 = 0 \\
\text{Subject to :} \quad & x_1 + x_2 + x_3 = 10 \\
& -x_1 + 4x_2 + x_4 = 16 \\
& 2x_1 + 5x_2 + x_5 = 30 \\
& x_1 - x_2 + x_6 = 6 \\
& x_1, x_2, x_3, x_4, x_5, x_6 \ge 0
\end{aligned}$$

Solving this problem using the simplex method the optimal tableau is :

$-z$	x_1	x_2	x_3	x_4	x_5	x_6	RHS
1	0	0	1/3	0	1/3	0	40/3
0	0	0	13/3	1	−5/3	0	28/3
0	0	1	−2/3	0	1/3	0	10/3
0	1	0	5/3	0	−1/3	0	20/3
0	0	0	−7/3	0	2/3	1	8/3

Hence, the optimal solution is $z = -40/3$, $x_1 = 20/3$, $x_2 = 10/3$, $x_3 = 0$, $x_4 = 28/3$, $x_5 = 0$ and $x_6 = 8/3$.

Use this information to find the ranges within which the requirements b_1, b_2, b_3 and b_4 can vary without producing a change in the optimal basis.

Solution
In this problem all constraints are of type 1.

Constraint (1)
The slack variable associated with this constraint is x_3. The optimal value of x_3 is 0. Hence, constraint (1) is tight at the optimum. The permissible increase in b_1 is :

$$\min\left\{\frac{b_i}{-a_{is}}\,\Big|\,a_{is}<0;1\le i\le m\right\}, \qquad \min\left\{\frac{b_i}{-a_{i3}}\,\Big|\,a_{i3}<0;1\le i\le 4\right\},$$

$$\min\left\{\frac{10/3}{-(-2/3)},\frac{8/3}{-(-7/3)}\right\}, \qquad \min\left\{\frac{10/3}{2/3},\frac{8/3}{7/3}\right\}, \qquad \min\{5,8/7\} \quad i.e. \ \ 8/7$$

The permissible decrease in b_1 is :

$$\min\left\{\frac{b_i}{a_{is}}\,\Big|\,a_{is}>0;1\le i\le m\right\}, \qquad \min\left\{\frac{b_i}{a_{i3}}\,\Big|\,a_{i3}>0;1\le i\le 4\right\},$$

$$\min\left\{\frac{28/3}{13/3},\frac{20/3}{5/3}\right\}, \qquad \min\left\{\frac{28}{13},4\right\} \quad i.e. \ \ 28/13$$

Hence, b_1 can take any value in the range $[10-28/13, 10+8/7]$ i.e. $[7.85, 11.14]$ without affecting the optimal basis.

Constraint (2)

The slack variable associated with this constraint is x_4. The optimal value of x_4 is $28/3$. Hence, constraint (2) is slack at the optimum. The permissible increase in b_2 is ∞ and the permissible decrease in b_2 is $28/3$. Hence, b_2 can take any value in the range $[16-28/3, 16+\infty)$ *i.e.* $[6.67, \infty)$ without affecting the optimal basis.

Constraint (3)

The slack variable associated with this constraint is x_5. The optimal value of x_5 is 0. Hence, constraint (3) is tight at the optimum. The permissible increase in b_3 is :

$$\min\left\{\frac{b_i}{-a_{is}}\,\middle|\,a_{is}<0; 1\leq i\leq m\right\}, \qquad \min\left\{\frac{b_i}{-a_{i5}}\,\middle|\,a_{i5}<0; 1\leq i\leq 4\right\},$$

$$\min\left\{\frac{28/3}{-(-5/3)}, \frac{20/3}{-(-1/3)}\right\}, \qquad \min\left\{\frac{28/3}{5/3}, \frac{20/3}{1/3}\right\}, \qquad \min\{28/5, 20\} \quad \textit{i.e.} \ \ 28/5$$

The permissible decrease in b_1 is :

$$\min\left\{\frac{b_i}{a_{is}}\,\middle|\,a_{is}>0; 1\leq i\leq m\right\}, \qquad \min\left\{\frac{b_i}{a_{i5}}\,\middle|\,a_{i5}>0; 1\leq i\leq 4\right\},$$

$$\min\left\{\frac{10/3}{1/3}, \frac{8/3}{2/3}\right\}, \qquad \min\{10, 4\} \quad \textit{i.e.} \ \ 4$$

Hence, b_3 can take any value in the range $[30-4, 30+28/5]$ *i.e.* $[26, 35.6]$ without affecting the optimal basis.

Constraint (4)

The slack variable associated with this constraint is x_6. The optimal value of x_6 is $8/3$. Hence, constraint (4) is slack at the optimum. The permissible increase in b_4 is ∞ and the permissible decrease in b_4 is $8/3$. Hence, b_4 can take any value in the range $[6-8/3, 6+\infty)$ *i.e.* $[3.33, \infty)$ without affecting the optimal basis.

Example 9.4

Consider the linear programming problem from Example 9.3. Find the new optimal value of z if :

(a) The value of b_1 changes to 7.95

(b) The value of b_2 changes to 16.23

(c) The value of b_3 changes to 37.84

(d) The value of b_4 changes to 6.36

Work to 2 decimal places.

Solution

The optimal value of z i.e. $z^* = -40/3 \approx -13.33$. The new optimal values of z i.e. z^+ can be calculated using the formula :

$$z^+ = z^* + \frac{\partial z^*}{\partial b_i} \delta b_i \quad ---- (1)$$

Negating the coefficients of the slack variables in the objective function row of the optimal simplex tableau, the optimal values of the dual variables are $y_1 = -1/3$, $y_2 = 0$, $y_3 = -1/3$ and $y_4 = 0$. Hence, by Theorem 6 :

$$\frac{\partial z^*}{\partial b_1} = -1/3 \approx -0.33, \quad \frac{\partial z^*}{\partial b_2} = 0, \quad \frac{\partial z^*}{\partial b_3} = -1/3 \approx -0.33 \text{ and } \frac{\partial z^*}{\partial b_4} = 0$$

(a) b_1 can take any value in the range $[7.85, 11.14]$ without affecting the optimal basis. The new value of b_1 i.e. 7.95 is <u>within</u> this range. Hence :

$$\delta b_1 = 7.95 - 10 \quad i.e. \quad \delta b_1 = -2.05$$

Substituting this into (1), the new optimal value of z is :

$$z^+ = z^* + \frac{\partial z^*}{\partial b_1} \delta b_1, \quad z^+ = -13.33 + (-0.33)(-2.05) \quad i.e. \quad z^+ \approx -12.65$$

(b) b_2 can take any value in the range $[6.67, \infty)$ without affecting the optimal basis. The new value of b_2 i.e. 16.23 is <u>within</u> this range. However, since $\frac{\partial z^*}{\partial b_2} = 0$ the change in the value of b_2 leaves the optimal value of z unchanged i.e. $z^+ \approx -13.33$.

(c) b_3 can take any value in the range $[26, 35.6]$ without affecting the optimal basis. The new value of b_3 i.e. 37.84 is <u>outside</u> this range. To find the new optimal value of z in this case the linear programming problem must be resolved.

(d) b_4 can take any value in the range $[3.33, \infty)$ without affecting the optimal basis. The new value of b_4 i.e. 6.36 is <u>within</u> this range. However, since $\frac{\partial z^*}{\partial b_4} = 0$ the change in the value of b_4 leaves the optimal value of z unchanged i.e. $z^+ \approx -13.33$.

Example 9.5
A furniture company manufactures two types of chair, a luxury model and a standard model using two types of skilled worker *i.e.* assemblers and finishers. Each luxury chair requires 1.5 hours of assembly time and 1 hour of finishing time. Each standard chair requires 0.5 hours of assembly time and 0.5 hours of finishing time. The company makes £20 profit on each luxury chair and £12 profit on each standard chair. Dues to other responsibilities within the company the assemblers are available for 100 hours only and the finishers are available for 80 hours only. The manager of the company is considering offering overtime to the workers but will do this only if it is in the company's interest.

(a) Formulate this problem as a linear programming model and solve it using the simplex method.

(b) Use the optimal solution obtained to determine the amount of overtime that should be offered to the workers and the maximum rate of pay.

Solution
(a) Let :

z be the total profit from manufacturing and selling the chairs.

x_1 be the number of luxury chairs produced.

x_2 be the number of standard chairs produced.

Then, the total profit is :
$$z = 20x_1 + 12x_2$$

The total profit is constrained by the availability of the workers. The assembler constraint is :
$$1.5x_1 + 0.5x_2 \leq 100$$

The finisher constraint is :
$$x_1 + 0.5x_2 \leq 80$$

Since the number of chairs produced cannot be negative, the total profit is also constrained by the trivial inequalities :
$$x_1, x_2 \geq 0$$

Combining these expressions the required linear programming model is :

$$\begin{aligned}
\text{Maximise} \quad & z = 20x_1 + 12x_2 \\
\text{Subject to :} \quad & 1.5x_1 + 0.5x_2 \leq 100 \quad \text{---- (1)} \\
& x_1 + 0.5x_2 \leq 80 \quad \text{---- (2)} \\
& x_1, x_2 \geq 0
\end{aligned}$$

Converting to a minimisation problem and adding slack variables x_3 and x_4 this problem can be written in canonical form as :

$$\text{Minimise} \qquad -\bar{z} - 20x_1 - 12x_2 = 0$$

$$\text{Subject to :} \qquad 1.5x_1 + 0.5x_2 + x_3 = 100$$

$$x_1 + 0.5x_2 + x_4 = 80$$

$$x_1, x_2, x_3, x_4 \geq 0$$

Solving this problem using the simplex method the optimal tableau is :

$-\bar{z}$	x_1	x_2	x_3	x_4	RHS
1	4	0	0	24	1920
0	0.5	0	1	-1	20
0	2	1	0	2	160

Hence, the optimal solution is $z = 1920$, $x_1 = 0$, $x_2 = 160$, $x_3 = 20$ and $x_4 = 0$ *i.e.* the optimal manufacturing strategy is to produce 0 luxury chairs, 160 standard chairs. The total profit is then £1920.

(b) The additional profits (per hour) the company can earn by increasing the hours worked by the assemblers and the finishers are given by the partial derivatives :

$$\frac{\partial z^*}{\partial b_1} \text{ and } \frac{\partial z^*}{\partial b_2}$$

respectively. By Theorem 6 the values of these partial derivatives are the optimal values of the dual variables y_1 and y_2. By Theorem 4 (for a maximisation problem) the optimal values of these dual variables are the coefficients of slack variables x_3 and x_4 in the objective function row of the optimal simplex tableau. Hence :

$$\frac{\partial z^*}{\partial b_1} = 0 \text{ and } \frac{\partial z^*}{\partial b_2} = 24$$

Since :

- $\frac{\partial z^*}{\partial b_1} = 0$, the company will not increase its profit by increasing the hours worked by the assemblers.

- $\frac{\partial z^*}{\partial b_2} = 24$, the company will increase its profit by £24 for each additional hour worked by the finishers.

Hence, it is in the company's interest to only increase the number of hours worked by the finishers.

The maximum increase in the number of hours worked by the finishers (without affecting the optimal basis) is the maximum permissible increase in the requirement b_2. Since the slack variable $x_4 = 0$, constraint (2) is tight at the optimum. Using the rule given earlier (for tight type 1 constraints), b_2 can be increased by at most :

$$\min\left\{\frac{b_i}{-a_{is}}\mid a_{is}<0; 1\leq i\leq m\right\}, \qquad \min\left\{\frac{b_i}{-a_{i4}}\mid a_{i4}<0; 1\leq i\leq 2\right\},$$

$$\min\left\{\frac{20}{-(-1)}\right\}, \qquad \min\left\{\frac{20}{1}\right\} \quad i.e. \;\; 20 \text{ hours}$$

Recommendation

The company should offer 20 hours overtime to the finishers. The maximum rate of pay should be less than £24 per hour.

Case 2 : Changes that <u>do</u> affect the optimal basis (and multiple changes)

Suppose that a requirement b_i changes by an amount that is outside of the permissible range or that several b_i change at the same time. Let $\underline{r}'$ be the new right-hand side vector in the first canonical form of the problem.

If the linear programming problem has been solved using one of the simplex methods the new optimal solution can be found by :

- Calculating the new right-hand side vector for the optimal tableau $\underline{r}^+$. This is done by premultiplying the vector $\underline{r}'$ by the optimal transformation matrix T i.e. $\underline{r}^+ = T\underline{r}'$.

  Note

  Changing the requirements does <u>not</u> affect the **optimality** of the solution since the optimality condition *i.e.* $(\underline{c}_N - \underline{c}_B B^{-1}N) \geq \underline{0}$ does not involve the b_i. However, changing the requirements <u>can</u> affect the **feasibility** of the solution since the elements of the new right-hand side vector $\underline{r}^+$ can be negative.

- Replacing the right-hand side vector in the optimal tableau with the new vector $\underline{r}^+$.

- Applying the dual simplex algorithm to the new tableau (if necessary) to re-establish feasibility.

If the linear programming problem has been solved using one of the revised simplex methods the new optimal solution can be found using a similar procedure to the one described above. However, here the dual simplex algorithm is applied to a simplified revised simplex tableau in the following form :

$-z/-\bar{z}$ + <u>Final</u> Basic Variables 　　　　　　　　　　　New Right-hand Side Vector *i.e.* $\underline{r}^+$

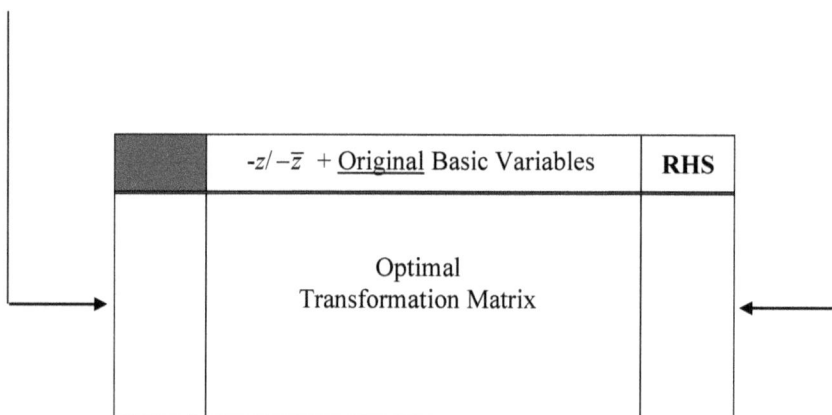

	$-z/-\bar{z}$ + <u>Original</u> Basic Variables	**RHS**
	Optimal Transformation Matrix	

Example 9.6

Consider the following linear programming problem :

$$\text{Minimise} \quad z = 2x_1 - x_2$$
$$\text{Subject to :} \quad x_2 \leq 1$$
$$x_1 \leq 1$$
$$x_1 - x_2 = 1/2$$
$$x_1 + x_2 \geq 1$$
$$x_1, x_2 \geq 0$$

This problem can be written in canonical form as :

$$\text{Minimise} \quad -z + 2x_1 - x_2 = 0$$
$$-w - 2x_1 + x_5 = -3/2$$
$$\text{Subject to :} \quad x_2 + x_3 = 1$$
$$x_1 + x_4 = 1$$
$$x_1 - x_2 + a_1 = 1/2$$
$$x_1 + x_2 - x_5 + a_2 = 1$$
$$x_1, x_2, x_3, x_4, x_5, a_1, a_2 \geq 0$$

The optimal two-phase simplex tableau is :

$-z$	$-w$	x_1	x_2	x_3	x_4	x_5	a_1	a_2	**RHS**
1	0	0	0	0	0	1/2	-3/2	-1/2	-5/4
0	1	0	0	0	0	0	1	1	0
0	0	0	0	1	0	1/2	1/2	-1/2	3/4
0	0	0	0	0	1	1/2	-1/2	-1/2	1/4
0	0	1	0	0	0	-1/2	1/2	1/2	3/4
0	0	0	1	0	0	-1/2	-1/2	1/2	1/4

Find the new optimal solution if the requirements change from $\begin{bmatrix} 1 & 1 & 1/2 & 1 \end{bmatrix}^{\mathrm{T}}$ to $\begin{bmatrix} 2 & 1 & 1 & 1/2 \end{bmatrix}^{\mathrm{T}}$.

Solution
From the optimal tableau :

$$B^{*-1} = \begin{bmatrix} 1 & 0 & 0 & 1 & 1 \\ 0 & 1 & 0 & 1/2 & -1/2 \\ 0 & 0 & 1 & -1/2 & -1/2 \\ 0 & 0 & 0 & 1/2 & 1/2 \\ 0 & 0 & 0 & -1/2 & 1/2 \end{bmatrix}$$

Reading the tableau from top to bottom, the optimal basis is $-w$, x_3, x_4, x_1 and x_2 in that order. Hence :

$$\underline{c}_B^* = \begin{bmatrix} c_{-w} & c_3 & c_4 & c_1 & c_2 \end{bmatrix} \quad i.e. \quad \underline{c}_B^* = \begin{bmatrix} 0 & 0 & 0 & 2 & -1 \end{bmatrix}$$

By definition :

$$T = \begin{bmatrix} 1 & -\underline{c}_B^* B^{*-1} \\ \underline{0} & B^{*-1} \end{bmatrix}$$

Here :

$$-\underline{c}_B^* B^{*-1} = -\begin{bmatrix} 0 & 0 & 0 & 2 & -1 \end{bmatrix}\begin{bmatrix} 1 & 0 & 0 & 1 & 1 \\ 0 & 1 & 0 & 1/2 & -1/2 \\ 0 & 0 & 1 & -1/2 & -1/2 \\ 0 & 0 & 0 & 1/2 & 1/2 \\ 0 & 0 & 0 & -1/2 & 1/2 \end{bmatrix} \quad i.e. \quad -\underline{c}_B^* B^{*-1} = \begin{bmatrix} 0 & 0 & 0 & -3/2 & -1/2 \end{bmatrix}$$

Substituting :

170

$$T = \begin{bmatrix} 1 & 0 & 0 & 0 & -3/2 & -1/2 \\ 0 & 1 & 0 & 0 & 1 & 1 \\ 0 & 0 & 1 & 0 & 1/2 & -1/2 \\ 0 & 0 & 0 & 1 & -1/2 & -1/2 \\ 0 & 0 & 0 & 0 & 1/2 & 1/2 \\ 0 & 0 & 0 & 0 & -1/2 & 1/2 \end{bmatrix}$$

In this example the new values of the requirements do not change the constant on the right-hand side of the w-row in the initial tableau. Hence, the new right-hand side vector is :

$$\underline{r}' = \begin{bmatrix} 0 \\ -3/2 \\ 2 \\ 1 \\ 1 \\ 1/2 \end{bmatrix}$$

Multiplying this vector by the optimal T matrix :

$$\underline{r}^+ = T\,\underline{r}', \qquad \underline{r}^+ = \begin{bmatrix} 1 & 0 & 0 & 0 & -3/2 & -1/2 \\ 0 & 1 & 0 & 0 & 1 & 1 \\ 0 & 0 & 1 & 0 & 1/2 & -1/2 \\ 0 & 0 & 0 & 1 & -1/2 & -1/2 \\ 0 & 0 & 0 & 0 & 1/2 & 1/2 \\ 0 & 0 & 0 & 0 & -1/2 & 1/2 \end{bmatrix} \begin{bmatrix} 0 \\ -3/2 \\ 2 \\ 1 \\ 1 \\ 1/2 \end{bmatrix} \qquad i.e. \quad \underline{r}^+ = \begin{bmatrix} -7/4 \\ 0 \\ 9/4 \\ 1/4 \\ 3/4 \\ -1/4 \end{bmatrix}$$

Replacing the right-hand side vector in the optimal tableau with the new vector $\underline{r}^+$:

$-z$	$-w$	x_1	x_2	x_3	x_4	x_5	a_1	a_2	RHS
1	0	0	0	0	0	1/2	−3/2	−1/2	−7/4
0	1	0	0	0	0	0	1	1	0
0	0	0	0	1	0	1/2	1/2	−1/2	9/4
0	0	0	0	0	1	1/2	−1/2	−1/2	1/4
0	0	1	0	0	0	−1/2	1/2	1/2	3/4
0	0	0	1	0	0	−1/2	−1/2	1/2	−1/4

This tableau is optimal but infeasible. The artificial variables can be discarded. The w-row and the $-w$ column then become trivial and can also be discarded. The tableau then becomes :

$-z$	x_1	x_2	x_3	x_4	x_5	RHS
1	0	0	0	0	1/2	$-7/4$
0	0	0	1	0	1/2	9/4
0	0	0	0	1	1/2	1/4
0	1	0	0	0	$-1/2$	3/4
0	0	1	0	0	$-1/2$	$-1/4$

Applying the dual simplex algorithm to this tableau using the greyed element in the x_5 column as pivot :

$-z$	x_1	x_2	x_3	x_4	x_5	RHS
1	0	1	0	0	0	-2
0	0	1	1	0	0	2
0	0	1	0	1	0	0
0	1	-1	0	0	0	1
0	0	-2	0	0	1	1/2

This tableau is now optimal and feasible. The new optimal solution is $-z = -2$ *i.e.* $z = 2$, $x_1 = 1$ and $x_2 = 0$.

Note
Changing the right-hand side vector has changed the optimal basis in this case *i.e.* the optimal basis with the original requirements was $\begin{bmatrix} x_3 & x_4 & x_1 & x_2 \end{bmatrix}^{\mathrm{T}}$. The optimal basis with the new requirements is $\begin{bmatrix} x_3 & x_4 & x_1 & x_5 \end{bmatrix}^{\mathrm{T}}$.

Example 9.7
Consider the following linear programming problem :

$$\text{Maximise} \quad z = 5x_1 + 3x_2$$
$$\text{Subject to :} \quad 4x_1 + 5x_2 \leq 1000$$
$$2x_1 + 2x_2 \leq 1000$$
$$3x_1 + 8x_2 \leq 1200$$
$$x_1, x_2 \geq 0$$

This problem can be written in canonical form as :

$$\text{Minimise} \quad -\bar{z} - 5x_1 - 3x_2 = 0$$
$$\text{Subject to :} \quad 4x_1 + 5x_2 + x_3 = 1000$$
$$2x_1 + 2x_2 + x_4 = 1000$$
$$3x_1 + 8x_2 + x_5 = 1200$$
$$x_1, x_2, x_3, x_4, x_5 \geq 0$$

The optimal transformation matrix and optimal basis are :

$$
\begin{bmatrix}
1 & 5/17 & 13/17 & 0 \\
0 & 5/17 & -4/17 & 0 \\
0 & -2/17 & 5/17 & 0 \\
0 & -2 & 1 & 1
\end{bmatrix}
\begin{bmatrix}
-\bar{z} \\
x_2 \\
x_1 \\
x_5
\end{bmatrix}
$$

Find the new optimal solution if the requirements change from $\begin{bmatrix} 1000 & 1000 & 1200 \end{bmatrix}^T$ to $\begin{bmatrix} 850 & 850 & 800 \end{bmatrix}^T$.

Solution

In this example $\underline{r}' = \begin{bmatrix} 0 \\ 850 \\ 850 \\ 800 \end{bmatrix}$. Hence :

$$
\underline{r}^+ = T\,\underline{r}', \qquad
\underline{r}^+ =
\begin{bmatrix}
1 & 5/17 & 13/17 & 0 \\
0 & 5/17 & -4/17 & 0 \\
0 & -2/17 & 5/17 & 0 \\
0 & -2 & 1 & 1
\end{bmatrix}
\begin{bmatrix}
0 \\
850 \\
850 \\
800
\end{bmatrix}
\quad i.e. \quad
\underline{r}^+ =
\begin{bmatrix}
900 \\
50 \\
150 \\
-50
\end{bmatrix}
$$

The revised simplex tableau is :

	$-\bar{z}$	x_3	x_4	x_5	**RHS**
$-\bar{z}$	1	5/17	13/17	0	900
x_2	0	5/17	$-4/17$	0	50
x_1	0	$-2/17$	5/17	0	150
x_5	0	-2	1	1	-50

This tableau is optimal but infeasible. Applying the dual simplex algorithm to this tableau using the greyed element in the x_3 column as pivot :

	$-\bar{z}$	x_3	x_4	x_5	**RHS**
$-\bar{z}$	1	0	31/34	5/34	15175/17
x_2	0	0	$-3/34$	5/34	725/17
x_1	0	0	4/17	$-1/17$	2600/17
x_3	0	1	$-1/2$	$-1/2$	25

173

This tableau is now optimal and feasible. The new optimal solution is $-\bar{z} = 15175/17$ *i.e.* $z = 15175/17$, $x_1 = 2600/17$ and $x_2 = 725/17$.

Note

Changing the right-hand side vector has changed the optimal basis in this case *i.e.* the optimal basis with the original requirements was $\begin{bmatrix} x_2 & x_1 & x_5 \end{bmatrix}^T$. The optimal basis with the new requirements is $\begin{bmatrix} x_2 & x_1 & x_3 \end{bmatrix}^T$.

9.3 Changes in the Vector $\underline{c}$

Suppose that the cost vector $\underline{c}$ changes to a new vector $\underline{c}'$. Let z^+ be the new value of z *i.e.* the value of z calculated by substituting the optimal values of the structural variables into the new objective function.

If the linear programming problem has been solved using one of the simplex methods the new optimal solution can be found by :

- Calculating the new (non-basic) z-row coefficients for the optimal tableau using the formula :

$$\underline{c}_N^+ = \underline{c}_N^* - \underline{c}_B^* B^{*-1} N^*$$

- Replacing z^* and $\underline{c}_N^*$ in the optimal tableau with the new values *i.e.* z^+ and $\underline{c}_N^+$.

- Applying the simplex algorithm to the new tableau (if necessary) to re-establish optimality.

If the linear programming problem has been solved using one of the revised simplex methods the new optimal solution can be found by :

- Calculating a new vector $-\underline{c}_B^* B^{*-1}$.

- Forming a new transformation matrix for the optimal tableau by replacing the appropriate part of the optimal transformation matrix with the new vector *i.e.*

$$T^+ = \begin{bmatrix} 1 & \boxed{-\underline{c}_B^* B^{*-1}} \\ \underline{0} & B^{*-1} \end{bmatrix} \quad \text{New vector}$$

- Replacing z^* and T^* in the optimal tableau with the new values *i.e.* z^+ and T^+.

- Applying the revised simplex algorithm (if necessary) to re-establish optimality.

Example 9.8
Consider the following linear programming problem :

$$\text{Minimise} \quad z = -2x_1 - 2x_2$$
$$\text{Subject to :} \quad -x_1 + x_2 \leq 2$$
$$x_1 + x_2 \leq 6$$
$$1/2 x_1 - x_2 \leq 3/2$$
$$x_1, x_2 \geq 0$$

This problem can be written in canonical form as :

$$\text{Minimise} \quad -z - 2x_1 - 2x_2 = 0$$
$$\text{Subject to :} \quad -x_1 + x_2 + x_3 = 2$$
$$x_1 + x_2 + x_4 = 6$$
$$1/2 x_1 - x_2 + x_5 = 3/2$$
$$x_1, x_2, x_3, x_4, x_5 \geq 0$$

The initial simplex tableau is :

(1)	(2)	(3)	(4)	(5)	(6)	(7)
$-z$	x_1	x_2	x_3	x_4	x_5	RHS
1	-2	-2	0	0	0	0
0	-1	1	1	0	0	2
0	1	1	0	1	0	6
0	1/2	-1	0	0	1	3/2

The optimal simplex tableau is :

(1)	(2)	(3)	(4)	(5)	(6)	(7)
$-z$	x_1	x_2	x_3	x_4	x_5	RHS
1	0	0	0	2	0	12
0	0	0	1	1/3	4/3	6
0	0	1	0	1/3	-2/3	1
0	1	0	0	2/3	2/3	5

Find the new optimal solution if the objective function changes from $z = -2x_1 - 2x_2$ to $z = -3x_1 - 5x_2$

.

Solution
The optimal solution of this problem is $-z = 12$ *i.e.* $z = -12$, $x_1 = 5$ and $x_2 = 1$.

Substituting the optimal values of the structural variables into the new objective function :

$$z^+ = -3 \times 5 - 5 \times 1 \quad \textit{i.e.} \quad z^+ = -20$$

If $z = -3x_1 - 5x_2$ then $\underline{c}' = \begin{bmatrix} -3 & -5 & 0 & 0 & 0 \end{bmatrix}$.

Reading the tableau from top to bottom, the optimal basis is x_3, x_2 and x_1 in that order. Hence :

$$\underline{c}_B^* = \begin{bmatrix} c_3 & c_2 & c_1 \end{bmatrix} \quad i.e. \quad \underline{c}_B^* = \begin{bmatrix} 0 & -5 & -3 \end{bmatrix}$$

$$\underline{c}_N^* = \begin{bmatrix} c_4 & c_5 \end{bmatrix} \quad i.e. \quad \underline{c}_N^* = \begin{bmatrix} 0 & 0 \end{bmatrix}$$

The basic variables in the initial tableau are x_3, x_4 and x_5. The order in which the columns associated with these variables must be arranged to form an identity matrix is (4), (5) and (6). Extracting the variable coefficients from columns (4), (5) and (6) from the optimal tableau :

$$B^{*-1} = \begin{bmatrix} 1 & 1/3 & 4/3 \\ 0 & 1/3 & -2/3 \\ 0 & 2/3 & 2/3 \end{bmatrix}$$

The non-basic variables in the optimal tableau are x_4 and x_5. Extracting the variable coefficients from columns (5) and (6) from the initial tableau :

$$N^* = \begin{bmatrix} 0 & 0 \\ 1 & 0 \\ 0 & 1 \end{bmatrix}$$

The coefficients of the non-basic variables are given by the formula :

$$\underline{c}_N^+ = \underline{c}_N^* - \underline{c}_B^* B^{*-1} N^*$$

Substituting :

$$\underline{c}_N^+ = \begin{bmatrix} 0 & 0 \end{bmatrix} - \begin{bmatrix} 0 & -5 & -3 \end{bmatrix} \begin{bmatrix} 1 & 1/3 & 4/3 \\ 0 & 1/3 & -2/3 \\ 0 & 2/3 & 2/3 \end{bmatrix} \begin{bmatrix} 0 & 0 \\ 1 & 0 \\ 0 & 1 \end{bmatrix} \quad i.e. \quad \underline{c}_N^+ = \begin{bmatrix} 11/3 & -4/3 \end{bmatrix}$$

Replacing z^* and $\underline{c}_N^*$ in the optimal tableau with the new values *i.e.* z^+ and $\underline{c}_N^+$ the new tableau becomes :

$-z$	x_1	x_2	x_3	x_4	x_5	**RHS**
1	0	0	0	11/3	−4/3	20
0	0	0	1	1/3	4/3	6
0	0	1	0	1/3	−2/3	1
0	1	0	0	2/3	2/3	5

This tableau is not optimal. Applying the simplex algorithm to this tableau using the greyed element in the x_5 column as pivot :

$-z$	x_1	x_2	x_3	x_4	x_5	RHS
1	0	0	1	4	0	26
0	0	0	3/4	1/4	1	9/2
0	0	1	1/2	1/2	0	4
0	1	0	-1/2	1/2	0	2

This tableau is optimal. The new optimal solution is $-z = 26$ *i.e.* $z = -26$, $x_1 = 2$ and $x_2 = 4$.

Example 9.9
Consider the following linear programming problem :

$$\begin{aligned} \text{Minimise} \quad & z = -5x_1 - 2x_2 - 2x_3 \\ \text{Subject to :} \quad & x_1 + x_2 + x_3 = 1 \\ & x_1 + 2x_2 - x_3 = 2 \\ & x_1, x_2, x_3 \geq 0 \end{aligned}$$

This problem can be written in canonical form as :

$$\begin{aligned} \text{Minimise} \quad & -z - 5x_1 - 2x_2 - 2x_3 = 0 \\ & -w - 2x_1 - 3x_2 = -3 \\ \text{Subject to :} \quad & x_1 + x_2 + x_3 + a_1 = 1 \\ & x_1 + 2x_2 - x_3 + a_2 = 2 \\ & x_1, x_2, x_3, a_1, a_2 \geq 0 \end{aligned}$$

The optimal transformation matrix, optimal basis and optimal right-hand side vector are :

$$\begin{bmatrix} 1 & 3 & 11 & 0 \\ 0 & 1 & 3 & 0 \\ 0 & -1 & -2 & 0 \\ 0 & 1 & 1 & 1 \end{bmatrix} \begin{bmatrix} -z \\ x_1 \\ x_2 \\ a_2 \end{bmatrix} \begin{bmatrix} 2 \\ 0 \\ 1 \\ 0 \end{bmatrix}$$

Find the new optimal solution if the objective function changes from $z = -5x_1 - 2x_2 - 2x_3$ to $z = -3x_1 - 4x_2 - x_3$.

Solution
Substituting the optimal values of the structural variables into the new objective function :

$$z^+ = -3 \times 0 - 4 \times 1 - 0 \quad \textit{i.e.} \quad z^+ = -4$$

The new right-hand side vector is :

$$\begin{bmatrix} -z \\ x_1 \\ x_2 \\ a_2 \end{bmatrix} = \begin{bmatrix} -(-4) \\ 0 \\ 1 \\ 0 \end{bmatrix} \quad i.e. \quad \begin{bmatrix} 4 \\ 0 \\ 1 \\ 0 \end{bmatrix}$$

In this example $\underline{c}_B^* = \begin{bmatrix} c_1 & c_2 & c_{a_2} \end{bmatrix}$ i.e. $\underline{c}_B^* = \begin{bmatrix} -3 & -4 & 0 \end{bmatrix}$

From the optimal transformation matrix :

$$B^{*-1} = \begin{bmatrix} 1 & 3 & 0 \\ -1 & -2 & 0 \\ 1 & 1 & 1 \end{bmatrix}$$

Now :

$$-\underline{c}_B^* B^{*-1} = -\begin{bmatrix} -3 & -4 & 0 \end{bmatrix}\begin{bmatrix} 1 & 3 & 0 \\ -1 & -2 & 0 \\ 1 & 1 & 1 \end{bmatrix} \quad i.e. \quad -\underline{c}_B^* B^{*-1} = \begin{bmatrix} -1 & 1 & 0 \end{bmatrix}$$

By definition :

$$T = \begin{bmatrix} 1 & -\underline{c}_B^* B^{*-1} \\ \underline{0} & B^{*-1} \end{bmatrix}$$

Substituting :

$$T = \begin{bmatrix} 1 & -1 & 1 & 0 \\ 0 & 1 & 3 & 0 \\ 0 & -1 & -2 & 0 \\ 0 & 1 & 1 & 1 \end{bmatrix}$$

The revised simplex tableau is :

	$-z$	$-w$	a_1	a_2	x_1	x_2	x_3	a_1	a_2	PC	RHS	Comments
$-z$	1	-1	1	0	-3	-4	-1		0		4	$c_3^{(6)} = 0$
x_1	0	1	3	0	-2	-3	0		0		0	a_1 discarded, a_2 ignored
x_2	0	-1	-2	0	1	1	1		0		1	Solution optimal
a_2	0	1	1	1	1	2	-1		1		0	

This tableau is optimal i.e. no further iterations are required. The new optimal solution is $-z = 4$ i.e. $z = -4$, $x_1 = 0$, $x_2 = 1$ and $x_3 = 0$.

Example 9.10
Consider the following linear programming problem :

$$\text{Minimise} \quad z = -x_1 - 2x_2$$
$$\text{Subject to :} \quad x_1 + x_2 \le 10$$
$$-x_1 + 4x_2 \le 16$$
$$2x_1 + 5x_2 \le 30$$
$$x_1 - x_2 \le 6$$
$$x_1, x_2 \ge 0$$

The optimal simplex tableau is :

$-z$	x_1	x_2	x_3	x_4	x_5	x_6	RHS
1	0	0	1/3	0	1/3	0	40/3
0	0	0	13/3	1	−5/3	0	28/3
0	0	1	−2/3	0	1/3	0	10/3
0	1	0	5/3	0	−1/3	0	20/3
0	0	0	−7/3	0	2/3	1	8/3

The optimal solution is $z = -40/3$, $x_1 = 20/3$ and $x_2 = 10/3$.

Find the range within which the coefficient of x_1 in the objective function can vary without the optimal solution being effected.

Solution
Let the coefficient of x_1 be c_1. Then, the objective function can be written as :

$$z = c_1 x_1 - 2x_2 \quad i.e. \quad -z + c_1 x_1 - 2x_2 = 0$$

In the optimal tableau x_1 and x_2 are basic variables *i.e.* the coefficients of x_1 and x_2 in the objective function row are zero. The coefficients of x_1 and x_2 in the new objective function can be made zero by :

- Adding $-c_1$ times the third constraint row to the objective function row.

- Adding twice the second constraint row to the objective function row *i.e.*

$$-z + c_1 x_1 - 2x_2 = 0 \qquad \text{(new objective function)}$$
$$-c_1 x_1 - 5/3 c_1 x_3 + 1/3 c_1 x_5 = -20/3 c_1 \qquad + \qquad (-c_1 \times \text{third constraint row})$$
$$2x_2 - 4/3 x_3 + 2/3 x_5 = 20/3 \qquad (2 \times \text{second constraint row})$$
___
$$-z + (-5/3 c_1 - 4/3)x_3 + (1/3 c_1 + 2/3)x_5 = -20/3(c_1 - 1)$$

For the solution to be optimal the coefficients of x_3 and x_5 on the left-hand side of this expression must be non-negative *i.e.*

$$-5/3c_1 - 4/3 \geq 0 \quad \text{---- (1)} \quad \underline{\text{and}} \quad 1/3c_1 + 2/3 \geq 0 \quad \text{---- (2)}$$

Solving these inequalities:

$$(1) \Rightarrow -5/3c_1 \geq 4/3 \quad i.e. \quad c_1 \leq \frac{3 \times 4}{-5 \times 3} \quad i.e. \quad c_1 \leq -4/5$$

$$(2) \Rightarrow 1/3c_1 \geq -2/3 \quad i.e. \quad c_1 \geq \frac{-2 \times 3}{3} \quad i.e. \quad c_1 \geq -2$$

Combining these expressions, the range within which the coefficient c_1 can vary without the optimal solution being effected is :

$$-2 \leq c_1 \leq -4/5$$

9.4 Changes in the Matrix A

Suppose that the coefficients of the structural variable x_i change. Let $\underline{a}'$ be the new column for x_i in the first canonical form of the problem.

If the linear programming problem has been solved using one of the simplex methods the new optimal solution can be found by :

- Calculating the new x_i column for the optimal tableau $\underline{a}^+$. This is done by premultiplying the vector $\underline{a}'$ by the optimal transformation matrix *i.e.* $\underline{a}^+ = T \underline{a}'$.

- Replacing the x_i column in the optimal tableau with the new vector $\underline{a}^+$.

- Performing further iterations of the simplex algorithm (if necessary) to re-establish optimality.

If the linear programming problem has been solved using one of the revised simplex methods the new optimal solution can be found by :

- Replacing the x_i column in the optimal tableau with the vector $\underline{a}'$.

- Performing further iterations of the revised simplex algorithm (if necessary) to re-establish optimality.

Example 9.11
Consider the following linear programming problem :

$$\begin{aligned}
\text{Minimise} \quad & z = 2x_1 - x_2 \\
\text{Subject to :} \quad & x_2 \le 1 \\
& x_1 \le 1 \\
& x_1 - x_2 = 1/2 \\
& x_1 + x_2 \ge 1 \\
& x_1, x_2 \ge 0
\end{aligned}$$

This problem can be written in canonical form as :

$$\begin{aligned}
\text{Minimise} \quad & -z + 2x_1 - x_2 = 0 \\
& -w - 2x_1 + x_5 = -3/2 \\
\text{Subject to :} \quad & x_2 + x_3 = 1 \\
& x_1 + x_4 = 1 \\
& x_1 - x_2 + a_1 = 1/2 \\
& x_1 + x_2 - x_5 + a_2 = 1 \\
& x_1, x_2, x_3, x_4, x_5, a_1, a_2 \ge 0
\end{aligned}$$

The optimal two-phase simplex tableau is :

$-z$	$-w$	x_1	x_2	x_3	x_4	x_5	a_1	a_2	**RHS**
1	0	0	0	0	0	1/2	$-3/2$	$-1/2$	$-5/4$
0	1	0	0	0	0	0	1	1	0
0	0	0	0	1	0	1/2	1/2	$-1/2$	3/4
0	0	0	0	0	1	1/2	$-1/2$	$-1/2$	1/4
0	0	1	0	0	0	$-1/2$	1/2	1/2	3/4
0	0	0	1	0	0	$-1/2$	$-1/2$	1/2	1/4

The optimal transformation matrix is :

$$T = \begin{bmatrix}
1 & 0 & 0 & 0 & -3/2 & -1/2 \\
0 & 1 & 0 & 0 & 1 & 1 \\
0 & 0 & 1 & 0 & 1/2 & -1/2 \\
0 & 0 & 0 & 1 & -1/2 & -1/2 \\
0 & 0 & 0 & 0 & 1/2 & 1/2 \\
0 & 0 & 0 & 0 & -1/2 & 1/2
\end{bmatrix}$$

Find the new optimal solution if the coefficient of x_2 in the third constraint changes from -1 to $-1/2$.

Solution

The new canonical form of the problem is :

$$\text{Minimise} \quad -z + 2x_1 - x_2 = 0$$
$$-w - 2x_1 - 1/2 x_2 + x_5 = -3/2$$
$$\text{Subject to :} \quad x_2 + x_3 = 1$$
$$x_1 + x_4 = 1$$
$$x_1 - 1/2 x_2 + a_1 = 1/2$$
$$x_1 + x_2 - x_5 + a_2 = 1$$
$$x_1, x_2, x_3, x_4, x_5, a_1, a_2 \geq 0$$

Hence, the new x_2 column is :

$$\begin{bmatrix} -1 \\ -1/2 \\ 1 \\ 0 \\ -1/2 \\ 1 \end{bmatrix}$$

Premultiplying this vector by the optimal transformation matrix the new x_2 column for the optimal tableau is :

$$\begin{bmatrix} 1 & 0 & 0 & 0 & -3/2 & -1/2 \\ 0 & 1 & 0 & 0 & 1 & 1 \\ 0 & 0 & 1 & 0 & 1/2 & -1/2 \\ 0 & 0 & 0 & 1 & -1/2 & -1/2 \\ 0 & 0 & 0 & 0 & 1/2 & 1/2 \\ 0 & 0 & 0 & 0 & -1/2 & 1/2 \end{bmatrix} \begin{bmatrix} -1 \\ -1/2 \\ 1 \\ 0 \\ -1/2 \\ 1 \end{bmatrix} \quad i.e. \quad \begin{bmatrix} -3/4 \\ 0 \\ 1/4 \\ -1/4 \\ 1/4 \\ 3/4 \end{bmatrix}$$

Replacing the x_2 column in the optimal tableau with the new one :

$-z$	$-w$	x_1	x_2	x_3	x_4	x_5	a_1	a_2	**RHS**
1	0	0	$-3/4$	0	0	$1/2$	$-3/2$	$-1/2$	$-5/4$
0	1	0	0	0	0	0	1	1	0
0	0	0	$1/4$	1	0	$1/2$	$1/2$	$-1/2$	$3/4$
0	0	0	$-1/4$	0	1	$1/2$	$-1/2$	$-1/2$	$1/4$
0	0	1	$1/4$	0	0	$-1/2$	$1/2$	$1/2$	$3/4$
0	0	0	$3/4$	0	0	$-1/2$	$-1/2$	$1/2$	$1/4$

This tableau is not optimal *i.e.* further iterations are required. The artificial variables can be discarded. The *w*-row and the $-w$ column then become trivial and can also be discarded. The tableauthen becomes :

$-z$	x_1	x_2	x_3	x_4	x_5	RHS
1	0	$-3/4$	0	0	$1/2$	$-5/4$
0	0	$1/4$	1	0	$1/2$	$3/4$
0	0	$-1/4$	0	1	$1/2$	$1/4$
0	1	$1/4$	0	0	$-1/2$	$3/4$
0	0	$3/4$	0	0	$-1/2$	$1/4$

Applying the simplex algorithm to this tableau using the greyed element in the x_2 column as pivot :

$-z$	x_1	x_2	x_3	x_4	x_5	RHS
1	0	0	0	0	0	-1
0	0	0	1	0	$2/3$	$2/3$
0	0	0	0	1	$1/3$	$1/3$
0	1	0	0	0	$-1/3$	$2/3$
0	0	1	0	0	$-2/3$	$1/3$

This tableau is now optimal. The new optimal solution is : $-z=-1$ *i.e.* $z=1$, $x_1=2/3$ and $x_2=1/3$.

Example 9.12
Consider the following linear programming problem :

$$\text{Minimise} \quad z=-5x_1-2x_2-2x_3$$
$$\text{Subject to :} \quad x_1+x_2+x_3=1$$
$$x_1+2x_2-x_3=2$$
$$x_1,x_2,x_3 \geq 0$$

This problem can be written in canonical form as :

$$\text{Minimise} \quad -z-5x_1-2x_2-2x_3=0$$
$$-w-2x_1-3x_2=-3$$
$$\text{Subject to :} \quad x_1+x_2+x_3+a_1=1$$
$$x_1+2x_2-x_3+a_2=2$$
$$x_1,x_2,x_3,a_1,a_2 \geq 0$$

The optimal transformation matrix, optimal basis and optimal right-hand side vector are :

$$\begin{bmatrix} 1 & 3 & 11 & 0 \\ 0 & 1 & 3 & 0 \\ 0 & -1 & -2 & 0 \\ 0 & 1 & 1 & 1 \end{bmatrix} \quad \begin{bmatrix} -z \\ x_1 \\ x_2 \\ a_2 \end{bmatrix} \quad \begin{bmatrix} 2 \\ 0 \\ 1 \\ 0 \end{bmatrix}$$

Find the new optimal solution if the coefficient of x_3 in the second constraint changes from -1 to $-1/2$.

Solution

The new canonical form of the problem is :

$$
\begin{aligned}
\text{Minimise} \quad & -z - 5x_1 - 2x_2 - 2x_3 = 0 \\
& -w - 2x_1 - 3x_2 - 1/2\,x_3 = -3 \\
\text{Subject to :} \quad & x_1 + x_2 + x_3 + a_1 = 1 \\
& x_1 + 2x_2 - 1/2\,x_3 + a_2 = 2 \\
& x_1, x_2, x_3, a_1, a_2 \geq 0
\end{aligned}
$$

Hence, the new x_3 column is :

$$
\begin{bmatrix} -2 \\ -1/2 \\ 1 \\ -1/2 \end{bmatrix}
$$

Replacing the old x_3 column with the new one, the revised simplex tableau is :

	$-z$	$-w$	a_1	a_2	x_1	x_2	x_3	a_1	a_2	PC	RHS	**Comments**
$-z$	1	3	11	0	-5	-2	-2		0		2	$c_3^{(6)} > 0$
x_1	0	1	3	0	-2	-3	$-1/2$		0		0	a_1 discarded, a_2 ignored
x_2	0	-1	-2	0	1	1	1		0		1	Solution optimal
a_2	0	1	1	1	1	2	$-1/2$		1		0	

This tableau is optimal *i.e.* no further iterations are required. The optimal solution is $-z = 2$ *i.e.* $z = -2$, $x_1 = 0$, $x_2 = 1$ and $x_3 = 0$.

9.5 Summary

Changes in the Vector $\underline{b}$

Case 1 : Changes that <u>do not</u> affect the optimal basis

Let x_s be the optimal value of the slack/surplus variable in the constraint.

Permissible Changes in b_i

Slack Constraints

      Type 1 : increase ∞, decrease x_s

      Type 3 : increase x_s, decrease ∞

Tight Constraints

Type 1 : increase $\min\left\{\dfrac{b_i}{-a_{is}}\mid a_{is}<0;1\le i\le m\right\}$, decrease $\min\left\{\dfrac{b_i}{a_{is}}\mid a_{is}>0;1\le i\le m\right\}$

Type 3 : increase $\min\left\{\dfrac{b_i}{a_{is}}\mid a_{is}>0;1\le i\le m\right\}$, decrease $\min\left\{\dfrac{b_i}{-a_{is}}\mid a_{is}<0;1\le i\le m\right\}$

The New Value of z

$$z^+ = z^* + \frac{\partial z^*}{\partial b_i}\delta b_i \quad \text{where } \frac{\partial z^*}{\partial b_i} \text{ is the optimal value of the } i\text{th dual variable.}$$

Case 2 : Changes that <u>do</u> affect the optimal basis (and multiple changes)

Write down the new right-hand side vector in the first canonical form of the problem *i.e.* $\underline{r}'$

- Calculate the new right-hand side vector for the optimal tableau using $\underline{r}^+ = T\,\underline{r}'$
- Substitute $\underline{r}^+$ into the optimal tableau and apply the dual simplex algorithm (if necessary) to re-establish feasibility.

Changes in the Vector $\underline{c}$

Calculate the new value of z *i.e.* z^+

The Simplex Method

- Calculate the new (non-basic) z-row coefficients for the optimal tableau using
$$\underline{c}_N^+ = \underline{c}_N^* - \underline{c}_B^* B^{*-1} N^*$$
- Substitute z^+ and $\underline{c}_N^+$ into the optimal tableau and apply the simplex algorithm (if necessary) to re-establish optimality.

The Revised Simplex Method

- Calculate $-\underline{c}_B^* B^{*-1}$ and form a new T matrix for the optimal tableau *i.e.* T^+
- Substitute z^+ and T^+ into the optimal tableau and apply the revised simplex algorithm (if necessary) to re-establish optimality.

Changes in the Matrix A

Write down the new variable column in the first canonical form of the problem *i.e.* $\underline{a}'$

The Simplex Method

- Calculate the new variable column for the optimal tableau using $\underline{a}^+ = T\,\underline{a}'$
- Substitute $\underline{a}^+$ into the optimal tableau and apply the simplex algorithm (if necessary) to re-establish optimality.

The Revised Simplex Method

- Substitute $\underline{a}'$ into the optimal tableau and apply the revised simplex algorithm (if necessary) to re-establish optimality.

9.6 Exercises 9

1. Consider the following linear programming problem :

   Minimise $\quad z = 3x_2 - x_3 + 8$

   Subject to : $\quad x_1 + x_2 + x_3 = 10$

   $\qquad\qquad 2x_1 + 3x_2 + x_3 = 15$

   $\qquad\qquad x_1, x_2, x_3 \geq 0$

   The optimal two-phase simplex tableau for this problem is :

$-z$	$-w$	x_1	x_2	x_3	a_1	a_2	RHS
1	0	0	2	0	2	-1	-3
0	1	0	0	0	1	1	0
0	0	0	-1	1	2	-1	5
0	0	1	2	0	-1	1	5

   Write down the matrix B^{*-1}.

2. Consider the following linear programming problem :

   Maximise $\quad z = 5x_1 + 2x_2 + 2x_3 - 7$

   Subject to : $\quad x_1 + x_2 + x_3 = 1$

   $\qquad\qquad x_1 + 2x_2 - x_3 = 2$

   $\qquad\qquad x_1, x_2, x_3 \geq 0$

   The optimal transformation matrix for this problem is :

   $$T = \begin{bmatrix} 1 & 0 & 0 & 0 \\ 0 & 1 & 3 & 0 \\ 0 & -1 & -2 & 0 \\ 0 & 1 & 1 & 1 \end{bmatrix}$$

   Write down the matrix B^{*-1}.

3. Consider the following linear programming problem :

Minimise $\quad z = 2x_1 - x_2$

Subject to : $\quad x_2 \le 1 \quad$ ---- (1)

$\qquad\qquad x_1 \le 1 \quad$ ---- (2)

$\qquad\qquad x_1 - x_2 = 1/2 \quad$ ---- (3)

$\qquad\qquad x_1 + x_2 \ge 1 \quad$ ---- (4)

$\qquad\qquad x_1, x_2 \ge 0$

The optimal two-phase simplex tableau for this problem is :

$-z$	$-w$	x_1	x_2	x_3	x_4	x_5	a_1	a_2	RHS
1	0	0	0	0	0	$1/2$	$-3/2$	$-1/2$	$-5/4$
0	1	0	0	0	0	0	1	1	0
0	0	0	0	1	0	$1/2$	$1/2$	$-1/2$	$3/4$
0	0	0	0	0	1	$1/2$	$-1/2$	$-1/2$	$1/4$
0	0	1	0	0	0	$-1/2$	$1/2$	$1/2$	$3/4$
0	0	0	1	0	0	$-1/2$	$-1/2$	$1/2$	$1/4$

Here, x_3 is the slack variable used in constraint (1), x_4 is the slack variable used in constraint (2), a_1 is the artificial variable used in constraint (3), x_5 is the surplus variable used in constraint (4) and a_2 is the artificial variable used in constraint (4).

(i) Use the tableau given to write down the optimal solution of this problem.

(ii) Find the ranges within which the requirements b_1, b_2, b_3 and b_4 can vary without producing a change in the optimal basis.

(iii) Find the new optimal value of z if :

(a) The value of b_1 changes to $7/4$.

(b) The value of b_2 changes to $5/4$.

(c) The value of b_3 changes to $3/8$.

(d) The value of b_4 changes to $3/4$.

4. Consider the following linear programming problem :

Maximise $z = x_1 + x_2$

Subject to : $4x_1 + 5x_2 \leq 200$

$x_1 + 4x_2 \geq 40$

$2x_1 + 3x_2 = 90$

$x_1, x_2 \geq 0$

The optimal two-phase simplex tableau for this problem is :

$-\bar{z}$	$-w$	x_1	x_2	x_3	x_4	a_1	a_2	RHS
1	0	0	1/2	0	0	0	0.5	45
0	1	0	0	0	0	1	1	0
0	0	0	-1	1	0	0	-2	20
0	0	1	3/2	0	0	0	1/2	45
0	0	0	-5/2	0	1	-1	1/2	5

(i) Find the optimal transformation matrix for this problem.

(ii) Find the new optimal solution if the requirements change from $\begin{bmatrix} 200 & 40 & 90 \end{bmatrix}^T$ to $\begin{bmatrix} 220 & 50 & 100 \end{bmatrix}^T$.

5. Consider the following linear programming problem :

Maximise $z = 4x_1 + 3x_2$

Subject to : $x_1 + x_2 \leq 40$

$2x_1 + x_2 \leq 50$

$x_1, x_2 \geq 0$

The optimal simplex tableau is :

$-\bar{z}$	x_1	x_2	x_3	x_4	RHS
1	0	0	2	1	130
0	0	1	2	-1	30
0	1	0	-1	1	10

Find the new optimal solution if the objective function changes from $z = 4x_1 + 3x_2$ to $z = 2x_1 + 5x_2$.

6. Consider the following linear programming problem :

Maximise $\quad z = x_1 + x_2 - 3/2$
Subject to : $\quad x_1 - x_2 \geq -1/2$
$\quad\quad\quad x_1 \leq 3$
$\quad\quad\quad x_1, x_2 \geq 0$

The optimal transformation matrix, optimal basis and optimal right-hand side vector are :

$$\begin{bmatrix} 1 & 1 & 2 \\ 0 & 1 & 1 \\ 0 & 0 & 1 \end{bmatrix} \quad \begin{bmatrix} -\bar{z} \\ x_2 \\ x_1 \end{bmatrix} \quad \begin{bmatrix} 5 \\ 7/2 \\ 3 \end{bmatrix}$$

Find the new optimal solution if the objective function changes from $z = x_1 + x_2 - 3/2$ to $z = 3x_1 + 4x_2 - 3/2$.

7. Consider the following linear programming problem :

Minimise $\quad z = -x_1 - 2x_2$
Subject to : $\quad x_1 + x_2 \leq 10$
$\quad\quad\quad -x_1 + 4x_2 \leq 16$
$\quad\quad\quad 2x_1 + 5x_2 \leq 30$
$\quad\quad\quad x_1 - x_2 \leq 6$
$\quad\quad\quad x_1, x_2 \geq 0$

The optimal simplex tableau is :

$-z$	x_1	x_2	x_3	x_4	x_5	x_6	RHS
1	0	0	1/3	0	1/3	0	40/3
0	0	0	13/3	1	−5/3	0	28/3
0	0	1	−2/3	0	1/3	0	10/3
0	1	0	5/3	0	−1/3	0	20/3
0	0	0	−7/3	0	2/3	1	8/3

The optimal solution is $z = -40/3$, $x_1 = 20/3$ and $x_2 = 10/3$.

Find the range within which the coefficient of x_2 in the objective function can vary without the optimal solution being effected.

8. Consider the following linear programming problem :

Maximise $\quad z = 0.4x_1 + 0.25x_2$

Subject to : $\quad 3x_1 + 6x_2 \leq 1100$

$\qquad\qquad 4x_1 + 7x_2 \leq 1900$

$\qquad\qquad 5x_1 + 4x_2 \leq 1400$

$\qquad\qquad x_1, x_2 \geq 0$

The optimal simplex tableau is :

$-\overline{z}$	x_1	x_2	x_3	x_4	x_5	RHS
1	0	0.07	0	0	0.08	112.00
0	0	3.60	1	0	-0.60	260.00
0	0	3.80	0	1	-0.80	780.00
0	1	0.80	0	0	0.20	280.00

Find the new optimal solution if the coefficient of x_2 in the third constraint changes from 4 to -10. Work to 2 decimal places.

9. Consider the following linear programming problem :

Maximise $\quad z = 25x_1 + 50x_2$

Subject to : $\quad 5x_1 + 14x_2 \leq 4200$

$\qquad\qquad 15x_1 + 14x_2 \leq 6300$

$\qquad\qquad 15x_1 + 7x_2 \leq 5250$

$\qquad\qquad x_1, x_2 \geq 0$

The optimal transformation matrix, optimal basis and optimal right-hand side vector are :

$$\begin{bmatrix} 1.0000 & 2.8550 & 0.7150 & 0.0000 \\ 0.0000 & 0.1071 & -0.0357 & 0.0000 \\ 0.0000 & -0.1000 & 0.1000 & 0.0000 \\ 0.0000 & -13.0001 & -1.2500 & 1.0000 \end{bmatrix} \begin{bmatrix} -\overline{z} \\ x_2 \\ x_1 \\ x_5 \end{bmatrix} \begin{bmatrix} 16500.3600 \\ 225.0135 \\ 209.9874 \\ 525.0945 \end{bmatrix}$$

Find the new optimal solution if the coefficient of x_1 in the second constraint changes from 15 to 8. Work to 4 decimal places.

10. Solving Linear Programming Problems Using the Solver Tool in *Excel*

10.1 Introduction
The spreadsheet package *Excel* provides a built-in tool called the **Solver** that can be used for solving integer and non-integer constrained linear programming problems and for performing a sensitivity analysis. This involves investigating the effects of small changes in the parameters of a linear programming problem *i.e.* is a form of post-optimal analysis.

10.2 Worksheet Design and Example
To solve a linear programming problem using the Solver a worksheet must be created that contains :

- Cells to hold the values of the structural variables. These cells are given an initial value *e.g.* zero and are then **changed** by the Solver to obtain the optimal solution.

- A cell containing a formula to evaluate the value of the objective function.

- Cells containing formulae to evaluate the expressions on the left-hand sides of the constraints.

- Cells containing the values on the right-hand sides of the constraints.

For example, consider the following linear programming problem :

$$\text{Minimise} \quad z = 4x_1 + 5x_2 + 6x_3$$
$$\text{Subject to :} \quad x_1 + x_2 \geq 11 \quad \text{---- (1)}$$
$$x_1 - x_2 \leq 5 \quad \text{---- (2)}$$
$$-x_1 - x_2 + x_3 = 0 \quad \text{---- (3)}$$
$$x_1, x_2, x_3 \geq 0$$

A suitable worksheet for solving this problem using the Solver is shown below :

	A	B	C
1	**Solver Example**		
2			
3	Variables		
4	z : 0		
5	$x1$: 0		
6	$x2$: 0		
7	$x3$: 0		
8			
9	Constraints		
10		LHS	RHS
11	(1) :	0	11
12	(2) :	0	5
13	(3) :	0	0

Here :

- Cells B5, B6 and B7 contain the values of the structural variables. Here, these cells are given an initial value of zero. However, other initial values could be used.

- Cell B4 contains a formula to evaluate the objective function *i.e.* the expression $4x_1 + 5x_2 + 6x_3$. The values of x_1, x_2 and x_3 in the formula are taken from cells B5, B6 and B7 respectively. Hence, in *Excel* syntax this formula is = 4*B5+5*B6+6*B7

- Cells B11, B12 and B13 contain formulae to evaluate the expressions on the left-hand sides of the constraints. For example, cell B11 contains a formula to evaluate the expression $x_1 + x_2$. In *Excel* syntax this formula is = B5+B6

- Cells C11, C12 and C13 contain the values on the right-hand sides of the constraints.

The Solver command is usually found in the **Tools** menu. The main Solver window allows the parameters of the model to be entered *e.g.*

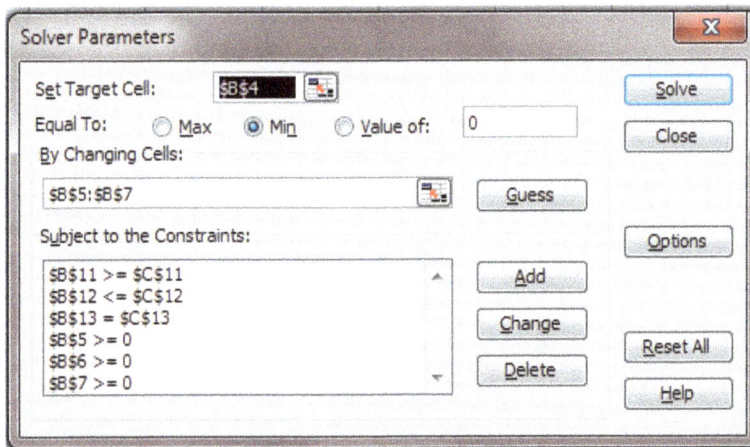

The constraints can be entered by clicking the **Add...** button *e.g.*

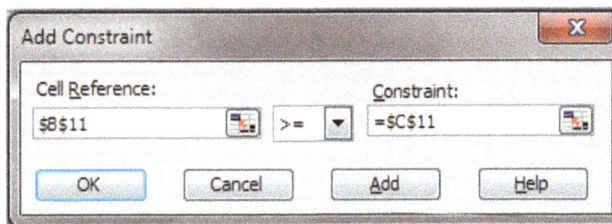

The down arrow button in the centre of this window provides access to the various comparison operators *i.e.* $\leq, \geq, =$. The **Add** button stores the constraint that has been entered and then repositions the cursor in the **Cell Reference** field so that another constraint can be added. The **OK** button returns control to the main Solver window.

Excel will find the optimal solution more quickly if the problem is defined to be a **linear model**. This is done by :

- Clicking the **Options...** button in the main Solver window *i.e.*

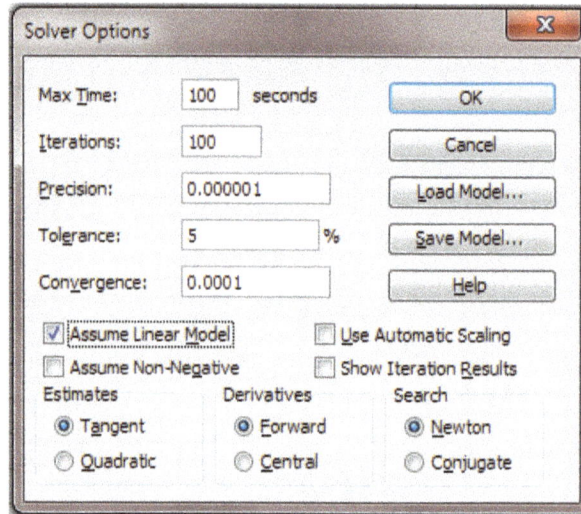

- Clicking in the **Assume Linear Model** check box and then clicking the **OK** button.

The optimal solution can be found by clicking the **Solve** button in the main Solver window. *Excel* then attempts to find a solution and displays an appropriate summary message *e.g.*

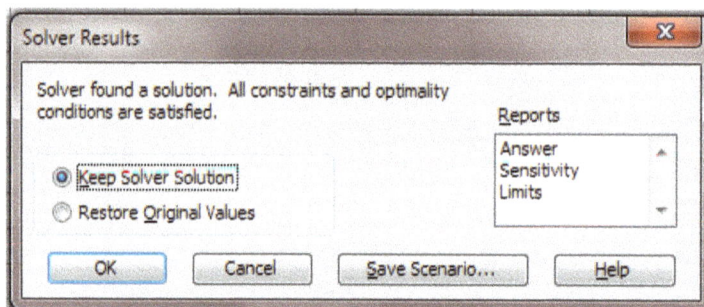

The optimal solution can be reflected into the worksheet by selecting the **Keep Solver Solution** option and clicking the **OK** button *e.g.*

	A	B	C
1	**Solver Example**		
2			
3	**Variables**		
4	$z :$	113	
5	$x1 :$	8	
6	$x2 :$	3	
7	$x3 :$	11	
8			
9	**Constraints**		
10		**LHS**	**RHS**
11	(1) :	11	11
12	(2) :	5	5
13	(3) :	0	0

It can be seen from this worksheet that the optimal solution of this linear programming problem is $z = 113$, $x_1 = 8$, $x_2 = 3$ and $x_3 = 11$.

10.3 Exercises 10a

Complete the following exercises from Chapter 2, Chapter 6 and Chapter 8 :

(i)   Exercises 2, Page 38, Question 3(ii)

(ii)   Exercises 6, Page 99, Question 4(i)

(iii) Exercises 6, Page 99, Question 5(i)

(iv)   Exercises 6, Page 99, Question 6(i)

(v)   Exercises 8, Page 155, Question 6(ii)

10.4 Engineering Application : Mixing Asphalt

Asphalt is made from bitumen, flux oil and aggregates such as granite, sand and limestone. The proportions of bitumen and flux oil are fixed and so engineers are concerned only with the mix of the aggregates used in the material. Each aggregate is graded in the laboratory by passing it through a series of sieves that have decreasing mesh sizes (*e.g.* from 20*mm* down to 75 *microns*). The sieves are stacked together with the largest sieve on top and the smallest sieve at the bottom *i.e.*

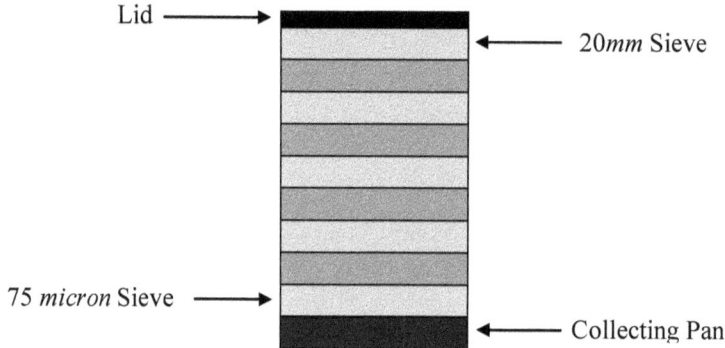

A sample of the aggregate is poured into the top sieve and the assembly is loaded into a machine that 'shakes' it for approximately 15 minutes. At the end of this time the amount of aggregate remaining in each sieve is weighed and the percentage passing through each sieve is calculated. This collection of percentages is called the **specification** of the aggregate. The aggregates are graded regularly as their specification can vary as they are extracted from different parts of the quarry.

The aggregate mix used in asphalt is controlled by British Standards to ensure that the asphalt has the required material properties. These standards define the percentage of the mix that must pass through each sieve in the set. For example, the specification for a 14*mm* Marshall Wearing Course (the asphalt used on aircraft runways) states that the percentages must lie within the following ranges :

	20*mm*	14*mm*	10*mm*	6.3*mm*	Sieve 3.35*mm*	2.36*mm*	425*mic*	150*mic*	75*mic*
Min %	100	86	78	66	52	34	19	9	3
Max %	100	100	90	79	65	49	33	17	6

The aggregates are purchased from a variety of companies and have different prices. Naturally, the suppliers of asphalt will wish to use the cheapest permissible mix. Hence, the aim is to determine the proportions in which to mix the aggregates so that the combined mix satisfies the British Standard and that the total cost of the mix is a minimum.

Mixing problems of this kind can be formulated and solved as linear programming models.

The Linear Programming Model

Let :

C be the total cost of the mix.

n be the number of materials used.

m be the number of sieves used.

x_i be the proportion of material i used in the mix. These proportions must be non-negative.

c_i be the price per tonne of material i.

P_{ij} be the percentage of material i that passes through sieve j.

L_j be the lower limit in the specification for sieve j i.e. the minimum percentage of the mix that must pass through sieve j.

U_j be the upper limit in the specification for sieve j i.e. the maximum percentage of the mix that must pass through sieve j.

Then, the cheapest mix of the aggregates that satisfies the British Standard specification can be found by minimising the objective function :

$$C = \sum_{i=1}^{n} c_i x_i \quad \text{---- (1)}$$

subject to the constraints :

$$\sum_{i=1}^{n} x_i = 1 \quad i.e. \quad 100 \sum_{i=1}^{n} x_i = 100 \quad \text{---- (2)}$$

$$\sum_{i=1}^{n} P_{ij} x_i \leq U_j \quad j = 1,...,m \quad \text{---- (3)}$$

$$\sum_{i=1}^{n} P_{ij} x_i \geq L_j \quad j = 1,...,m \quad \text{---- (4)}$$

and the non-negativity conditions :

$$x_i \geq 0 \quad i = 1,...,n \quad \text{---- (5)}$$

A linear programming model of this kind i.e. one in which the objective function coefficients c_i are all positive can be solved using the dual simplex algorithm. This is described in Chapter 7. Here, the optimal solution i.e. the mix that gives the minimum total cost will be found using the Solver.

The Spreadsheet Model

The spreadsheet model consists of four linked worksheets. The first worksheet is a database that lists the aggregates available to the company and gives details of their supplier and price *i.e.*

	A	B	C	D	E
1	**Material Information**				
2					
3	**Material Number**	**Description**	**Supplier**	**Price Per Tonne**	
4	1	14*mm* Granite	Roadstone Aggregates Ltd.	£7.93	
5	2	10*mm* Granite	Roadstone Aggregates Ltd.	£8.09	
6	3	6*mm* Granite	Roadstone Aggregates Ltd.	£8.47	
7	4	Course Sand	Hall Aggregates	£4.35	
8	5	Medium Sand	Beds Silica	£4.63	
9	6	Fine Sand	Kendall & Prestwick	£4.94	
10	7	Limestone Filler	Stevenson Minerals	£10.60	
11					

Figure 1

The second worksheet is a database that lists the material specifications *i.e.* the gradings obtained in the laboratory *i.e.*

	A	B	C	D	E	F	G	H	I	J	K
1	**Grading Information**										
2											
3	**Percentage of material passing through each sieve**										
4											
5		*--- Sieve Size ---*									
6	**Material Number**	**20.00mm**	**14.00mm**	**10.00mm**	**6.30mm**	**3.35mm**	**2.36mm**	**425mic**	**150mic**	**75mic**	
7	1	100.00	73.10	5.30	0.70	0.00	0.00	0.00	0.00	0.00	
8	2	100.00	100.00	72.30	3.40	0.00	0.00	0.00	0.00	0.00	
9	3	100.00	100.00	100.00	95.50	12.10	1.20	0.00	0.00	0.00	
10	4	100.00	100.00	100.00	100.00	97.90	63.00	34.70	13.00	4.00	
11	5	100.00	100.00	100.00	100.00	98.40	84.20	48.20	3.60	0.70	
12	6	100.00	100.00	100.00	100.00	100.00	100.00	100.00	53.20	4.90	
13	7	100.00	100.00	100.00	100.00	100.00	100.00	100.00	100.00	90.20	
14											

Figure 2

In a commercial system these databases will contain hundreds of entries. Wherever possible the other worksheets in the spreadsheet model extract information from these databases using appropriate external references. In this way any changes made to the data (*e.g.* to the material prices or to the specifications) are updated in the other worksheets automatically.

The third worksheet is used for calculating the optimal mix of the aggregates *i.e.*

	A	B	C	D	E	F	G	H	I	J	K	L
1	**Mix Information**											
2												
3			*-- Sieve Size --*									
4	**Material Number**	**Proportion**	**20.00mm**	**14.00mm**	**10.00mm**	**6.30mm**	**3.35mm**	**2.36mm**	**425mic**	**150mic**	**75mic**	
5	1	10.00%	10.00	7.31	0.53	0.07	0.00	0.00	0.00	0.00	0.00	
6	2	10.00%	10.00	10.00	7.23	0.34	0.00	0.00	0.00	0.00	0.00	
7	3	20.00%	20.00	20.00	20.00	19.10	2.42	0.24	0.00	0.00	0.00	
8	4	10.00%	10.00	10.00	10.00	10.00	9.79	6.30	3.47	1.30	0.40	
9	5	20.00%	20.00	20.00	20.00	20.00	19.68	16.84	9.64	0.72	0.14	
10	6	20.00%	20.00	20.00	20.00	20.00	20.00	20.00	20.00	10.64	0.98	
11	7	10.00%	10.00	10.00	10.00	10.00	10.00	10.00	10.00	10.00	9.02	
12												
13	**Total % :**	100.00	100.00	97.31	87.76	79.51	61.89	53.38	43.11	22.66	10.54	
14												
15			**Material Specification (*i.e.* % passing through each sieve)**									
16		**Min %**	100	86	78	66	52	34	19	9	3	
17		**Max %**	100	100	90	79	65	49	33	17	6	
18												
19		**Cost of Mix :**	£6.71	**Per Tonne (Excluding plant mixing cost)**								
20												

Figure 3

The material proportions in cells B5:B11 are initial estimates. Cells C5:K11 contain formulae for calculating the contribution made by each aggregate to the percentage of the mix that passes through each sieve. For example, the value 10.00 in cell C5 indicates that material 1 (*i.e.* 14mm granite) contributes 10% to the percentage of the mix that passes through the 20mm sieve. The formula for calculating this value is :

$$= \text{Sheet2!B7*\$B\$5}$$

where Sheet2!B7 is the percentage of 14mm granite passing through the 20mm sieve in the material specification and \$B\$5 is the proportion of 14mm granite being used in the mix. Once the formulae in cells C5:C11 have been created the remaining formulae in each row can be set-up quickly by using the **Edit:Fill:Right** command.

Cells C5:K5 contain the percentage of the aggregate mix that passes through each sieve (*i.e.* the sum of the material contributions) and cells B16:K17 contain the British Standard specification for the mix.

The cost of mix value in cell C19 is calculated using the formula :

$$= \text{Sheet1!D4*B5+ Sheet1!D5*B6+ Sheet1!D6*B7+ Sheet1!D7*B8+ Sheet1!D8*B9+}$$
$$\text{Sheet1!D9*B10+ Sheet1!D10*B11}$$

where Sheet1!D4:Sheet1!D10 are the material prices and B5:B11 are the material proportions being used in the mix.

It may not be obvious how the worksheet shown in Figure 3 relates to the linear programming model given earlier. However :

- The total cost of the mix C is given in cell C19.

- The x_i values are the material proportions in cells B5:B11. These are the values that must be varied in order to find the optimal solution.

- The value of $100 \sum_{i=1}^{n} x_i$ is given in cell B13.

- The $\sum_{i=1}^{n} P_{ij} x_i$ values are the totals in cells C13:K13.

- The U_j values are the upper limits in cells C17:K17.

- The L_j values are the lower limits in cells C16:K16.

Hence, the mathematical model can be rewritten using absolute *Excel* cell references as :

$$\text{Minimise } \$C\$19 \quad \text{---- (1)}$$

subject to the constraints :

$$\$B\$13 = 100 \quad \text{---- (2)}$$

$$\left. \begin{array}{l} \$C\$13 \leq \$C\$17 \\ \$D\$13 \leq \$D\$17 \\ \cdot \\ \cdot \\ \cdot \\ \$K\$13 \leq \$K\$17 \end{array} \right\} \quad \text{---- (3)}$$

$$\left. \begin{array}{l} \$C\$13 \geq \$C\$16 \\ \$D\$13 \geq \$D\$16 \\ \cdot \\ \cdot \\ \cdot \\ \$K\$13 \geq \$K\$16 \end{array} \right\} \quad \text{---- (4)}$$

and the non-negativity conditions *i.e.*

$$\left. \begin{array}{l} \$B\$5 \geq 0 \\ \$B\$6 \geq 0 \\ \cdot \\ \cdot \\ \cdot \\ \$B\$11 \geq 0 \end{array} \right\} \quad \text{---- (5)}$$

The model is entered into the Solver in the same way as in the earlier example. The main Solver window for this model is shown below :

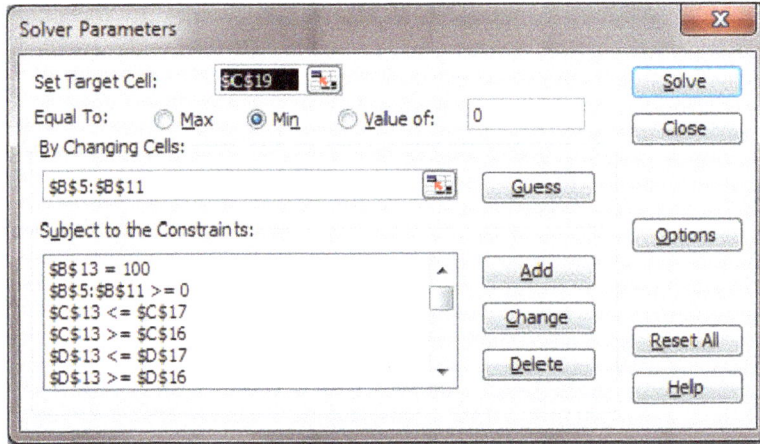

Defining the problem to be a **linear model** and clicking the **Solve** button in the main Solver window produces the window below :

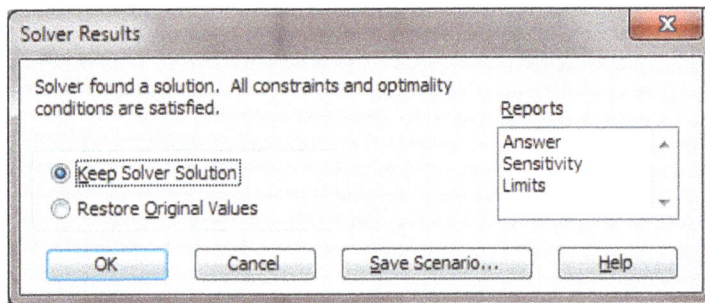

The optimal solution is then reflected into the worksheet by selecting the **Keep Solver Solution** option and clicking the **OK** button *i.e.*

	A	B	C	D	E	F	G	H	I	J	K	L
1	**Mix Information**											
2												
3			*--- Sieve Size ---*									
4	**Material Number**	**Proportion**	**20.00mm**	**14.00mm**	**10.00mm**	**6.30mm**	**3.35mm**	**2.36mm**	**425mic**	**150mic**	**75mic**	
5	1	18.94%	18.94	13.84	1.00	0.13	0.00	0.00	0.00	0.00	0.00	
6	2	14.68%	14.68	14.68	10.61	0.50	0.00	0.00	0.00	0.00	0.00	
7	3	0.00%	0.00	0.00	0.00	0.00	0.00	0.00	0.00	0.00	0.00	
8	4	65.93%	65.93	65.93	65.93	65.93	64.54	41.54	22.88	8.57	2.64	
9	5	0.00%	0.00	0.00	0.00	0.00	0.00	0.00	0.00	0.00	0.00	
10	6	0.06%	0.06	0.06	0.06	0.06	0.06	0.06	0.06	0.03	0.00	
11	7	0.40%	0.40	0.40	0.40	0.40	0.40	0.40	0.40	0.40	0.36	
12												
13	**Total % :**	100.00	100.00	94.91	78.00	67.02	65.00	41.99	23.33	9.00	3.00	
14												
15			**Material Specification (*i.e.* % passing through each sieve)**									
16		**Min %**	100	86	78	66	52	34	19	9	3	
17		**Max %**	100	100	90	79	65	49	33	17	6	
18												
19		**Cost of Mix :**	£5.60	**Per Tonne (Excluding plant mixing cost)**								
20												

Figure 4

The final worksheet in the spreadsheet model provides a summary of the optimal solution for the Technical Manager and the operators at the mixing plants.

	A	B	C	D	E	F	G	H	I
1	**Summary Report**								
2									
3		**Customer :**	RAF Conninsby						
4		**Material Description :**	14*mm* Marshall Wearing Course						
5		**Date :**	Tuesday 4th October 1955						
6		**Cost of Mix :**	£5.60						
7		**Total Tonnes Required :**	60,000						
8									
9	**Material Number**	**Description**	**Supplier**	**Price Per Tonne**	**Proportion**	**Kg Per Tonne**	**Tonnes Required**	**Cost**	
10	1	14*mm* Granite	Roadstone Aggregates Ltd.	£7.93	18.94%	189.38	11362.84	£90,107.34	
11	2	10*mm* Granite	Roadstone Aggregates Ltd.	£8.09	14.68%	146.77	8806.45	£71,244.21	
12	3	6*mm* Granite	Roadstone Aggregates Ltd.	£8.47	0.00%	0.00	0.00	£0.00	
13	4	Course Sand	Hall Aggregates	£4.35	65.93%	659.29	39557.29	£172,074.20	
14	5	Medium Sand	Beds Silica	£4.63	0.00%	0.00	0.00	£0.00	
15	6	Fine Sand	Kendall & Prestwick	£4.94	0.06%	0.56	33.89	£167.44	
16	7	Limestone Filler	Stevenson Minerals	£10.60	0.40%	3.99	239.52	£2,538.92	
17									
18								**Total :**	£336,132.12	
19									

Figure 5

The **Kg Per Tonne** values in cells F10:F16 are calculated by multiplying the material proportions by 1000. The **Tonnes Required** values in cells G10:G16 are calculated by multiplying the **Total Tonnes Required** value in cell C7 by the material proportions and the **Cost** values in cells H10:H16 are calculated by multiplying the **Tonnes Required** values in the previous column by the material prices.

Conclusion
Although the Solver has found the optimal solution of this mixing problem it can be unreliable on other optimisation problems *i.e.* it will sometimes fail to find a solution even when a feasible and optimal solution exists.

10.5 Later Versions of *Excel*
In later versions of *Excel* the Solver windows may be slightly different in appearance from the ones shown in this chapter. However, they will still provide the same tools, options and facilities.

10.6 Exercises 10b

The owners of Silverstone motor racing circuit wish to resurface Beckett's Corner using a special 10*mm* asphalt wearing course. The British Standard specification for this asphalt is shown in the table below.

				Sieve				
	14mm	**10mm**	**6.3mm**	**3.35mm**	**2.36mm**	**425mic**	**150mic**	**75mic**
Min %	100	82	69	51	32	15	7	2
Max %	100	100	83	68	48	29	11	5

The contractors hired to complete the work estimate that 5000 tonnes of this asphalt will be required. They intend to produce the aggregate mix using materials 2,3,4,5,6 and 7 in the databases shown in Figure 1 and Figure 2.

1. Open an *Excel* workbook and set up worksheets (similar to those shown in Figure 1, Figure 2 and Figure 3) to calculate the optimal blend of the aggregate mix.

2. Adjust the material proportions in your worksheet for calculating the optimal blend until the aggregate mix conforms to the British Standard specification shown in the table above.

3. Use the Solver to find the optimal blend of the aggregates used in the mix.

4. Set up a worksheet (similar to the one shown in Figure 5) to provide a summary of the optimal solution for the Technical Manager and the operators at the mixing plants.

Part Two - Assignment Problems

11. Solving Assignment Problems

11.1 Introduction
An assignment problem is one in which one set of quantities *e.g.* jobs must be assigned to another set of quantities *e.g.* machines in the optimal way. Unless stated otherwise the optimal assignment is the one that **minimises** the total cost or time *etc.* To illustrate how problems of this kind are solved consider the following example.

The Computer Centre Problem
In the Computer Centre of a large company there are five computers, C1 to C5 on which it is required to run five programs, P1 to P5. The running time (in minutes) of each program on each computer is shown in the table below :

		Program				
		P1	P2	P3	P4	P5
Computer	C1	5	8	5	7	9
	C2	10	8	3	8	5
	C3	7	10	5	4	6
	C4	9	9	17	12	9
	C5	8	6	3	10	8

The Computer Centre Manager wishes to find the assignment of programs to computers that will minimise the total running time.

Solving an Assignment Problem
An assignment problem can be formulated as a linear programming model. For example, consider the problem of assigning n jobs to n machines (one job to each machine) in the way that minimises the total cost. Suppose that :

- The total cost is z.

- The cost of assigning job i to machine j is c_{ij}.

- $x_{ij} = \begin{cases} 0 & \text{if the } i\text{th job is \textbf{NOT} assigned to the } j\text{th machine} \\ 1 & \text{if the } i\text{th job \textbf{IS} assigned to the } j\text{th machine} \end{cases}$

Then, the optimal assignment can be found by solving the following linear programming model :

$$\text{Minimise} \quad z = \sum_{i=1}^{n} \sum_{j=1}^{n} c_{ij} x_{ij}$$

$$\text{Subject to :} \quad \sum_{j=1}^{n} x_{ij} = 1 \quad i = 1, 2, \ldots, n \quad \textit{i.e. all machines must be used.}$$

$$\sum_{i=1}^{n} x_{ij} = 1 \quad j = 1, 2, \ldots, n \quad \textit{i.e. all jobs must be assigned.}$$

$$x_{ij} \geq 0 \quad \forall i, j$$

Solving an assignment problem in this way can be problematic. For example :

- The linear programming model above has $n \times n$ variables and $2n$ constraints *i.e.* even for a relatively small value of n this model will be very large. For example, suppose that a company wishes to assign 100 workers to 100 jobs. The linear programming model of this assignment problem will have 100x100 *i.e.* 10000 variables and 200 constraints.

- Linear programming models of this kind are highly **degenerate** *i.e.* conventional methods of solution such as the two-phase simplex method often experience cycling and fail to find the optimal solution.

11.2 The Hungarian Algorithm
To overcome these problems a special procedure has been developed called the **Hungarian algorithm**. This takes advantage of the special structure and sparsity of an assignment problem, it can overcome degeneracy and is more computationally efficient.

Steps in the Hungarian Algorithm

Step 1 : Form the initial matrix
The rows of the matrix represent one set of quantities and the columns of the matrix represent the other set of quantities. The elements of the matrix are the associated cost or times *etc.*

Example 11.1
The initial matrix for the Computer Centre problem is :

$$\begin{bmatrix} 5 & 8 & 5 & 7 & 9 \\ 10 & 8 & 3 & 8 & 5 \\ 7 & 10 & 5 & 4 & 6 \\ 9 & 9 & 17 & 12 & 9 \\ 8 & 6 & 3 & 10 & 8 \end{bmatrix}$$

Step 2 : Put a zero in each row and column
To do this :

1) Find any rows that do not contain a zero.

2) Identify the smallest element in each of these rows.

3) Form a new matrix by subtracting the smallest elements from the other elements in their respective rows.

4) Find any columns that do not contain a zero.

5) Identify the smallest element in each of these columns.

6) Form a new matrix by subtracting the smallest elements from the other elements in their respective columns.

Example 11.2

Applying this procedure to the initial matrix for the Computer Centre problem :

The smallest element in each row is :

$$\begin{bmatrix} 5^* & 8 & 5 & 7 & 9 \\ 10 & 8 & 3^* & 8 & 5 \\ 7 & 10 & 5 & 4^* & 6 \\ 9^* & 9 & 17 & 12 & 9 \\ 8 & 6 & 3^* & 10 & 8 \end{bmatrix}$$

Subtracting the starred elements from the other elements in their respective rows the new matrix is :

$$\begin{bmatrix} 0 & 3 & 0 & 2 & 4 \\ 7 & 5 & 0 & 5 & 2 \\ 3 & 6 & 1 & 0 & 2 \\ 0 & 0 & 8 & 3 & 0 \\ 5 & 3 & 0 & 7 & 5 \end{bmatrix}$$

All rows and columns now contain a zero *i.e.* no further subtractions are required.

Step 3 : Find the independent zeros
To do this :

1) Find a row that contains one (uncrossed out) zero only and circle that zero.

2) Cross out all other zeros in the column that contains the circled zero.

3) Repeat the steps 1) and 2) above until all the rows have been examined.

4) Repeat the steps 1), 2) and 3) above, this time finding columns that contain one (uncrossed out) zero only and crossing out the zeros in the appropriate rows.

Note : The steps above must be repeated until all zeros in the matrix are either circled or crossed out,

At the end of this process the **independent zeros** are the zeros that have been **circled**. Three possibilities exist:

- **There are n independent zeros**
  In this case the optimal assignment has been found and the algorithm **terminates**. The optimal assignment can be found by superimposing the positions of the independent zeros onto the original matrix. The optimal cost or time *etc.* can be found by finding the sum of the circled costs or times *etc.*

- **There are $> n$ independent zeros**
  In this case the optimal assignment is **non-unique**. The optimal assignment can be found by choosing any n **appropriate** independent zeros and then proceeding as above.

- **There are $< n$ independent zeros**
  In this case the optimal assignment has yet to be found and the algorithm proceeds to **Step 4**.

Example 11.3
Applying this procedure to the current matrix for the Computer Centre problem :

$$\begin{bmatrix} ⓪ & 3 & ⨉ & 2 & 4 \\ 7 & 5 & ⓪ & 5 & 2 \\ 3 & 6 & 1 & ⓪ & 2 \\ ⨉ & ⓪ & 8 & 3 & ⨉ \\ 5 & 3 & ⨉ & 7 & 5 \end{bmatrix}$$

Here there are 4 *(i.e. < n)* independent zeros *i.e.* the optimal assignment has yet to be found.

Step 4 : Find an improved assignment
To do this :

1) Draw onto the matrix the **minimum** number of horizontal and vertical lines needed to pass through **all** zeros. To do this :

- Find the row or column that contains the most zeros (without lines passing through them) and draw a line through it.

- If all zeros have a line passing through them **stop**. Otherwise repeat the step above.

   Note : If there are **m** independent zeros then this operation <u>cannot</u> be completed with less then **m** lines.

2) Find the smallest element in the matrix that does **not** have a line passing through it.

3) Form a new matrix by :

- Subtracting this element from all other elements that do **not** have a line passing through them.

- Adding this element to all elements that have **two** lines passing through them.

Example 11.4
Applying this procedure to the current matrix for the Computer Centre problem :

Drawing the lines :

Subtracting/adding the smallest element (*i.e.* 2) the new matrix is :

$$\begin{bmatrix} 0 & 3 & 2 & 4 & 4 \\ 5 & 3 & 0 & 5 & 0 \\ 1 & 4 & 1 & 0 & 0 \\ 0 & 0 & 10 & 5 & 0 \\ 3 & 1 & 0 & 7 & 3 \end{bmatrix}$$

Step 5 : If each row and column contains a zero go back to Step 3 otherwise go back to Step 2

At the next iteration **Step 3** produces the following matrix :

$$\begin{bmatrix} ⓪ & 3 & 2 & 4 & 4 \\ 5 & 3 & ✗ & 5 & ⓪ \\ 1 & 4 & 1 & ⓪ & ✗ \\ ✗ & ⓪ & 10 & 5 & ✗ \\ 3 & 1 & ⓪ & 7 & 3 \end{bmatrix}$$

Since there are 5 *(i.e. n)* independent zeros *i.e.* the optimal assignment has been found. Superimposing the positions of the independent zeros onto the original matrix :

$$\begin{bmatrix} ⑤ & 8 & 5 & 7 & 9 \\ 10 & 8 & 3 & 8 & ⑤ \\ 7 & 10 & 5 & ④ & 6 \\ 9 & ⑨ & 17 & 12 & 9 \\ 8 & 6 & ③ & 10 & 8 \end{bmatrix}$$

The optimal assignment is :

Computer	Program
C1	P1
C2	P5
C3	P4
C4	P2
C5	P3

The total running time is 5+5+4+9+3 = 26 minutes.

Accuracy Check
In **Step 2** and **Step 4** of the Hungarian algorithm values are subtracted from the elements of the matrix. The optimal assignment is correct if the sum of these subtractions is equal to the optimal cost or time *etc.*

For example, consider the Computer Centre problem. At **Step 2** the values 5,3,4,9,3 were subtracted from the rows of the matrix. At **Step 4** the value 2 was subtracted from the elements without a line passing through them. The sum of these subtractions is 5+3+4+9+3+2 = 26 = the total running time calculated. Hence, the optimal assignment calculated above is correct.

Summary of the Hungarian Algorithm

The Hungarian algorithm can be summarised as follows :

Step 1

Form the <u>initial matrix</u>. The rows of the matrix represent one set of quantities and the columns of the matrix represent the other set of quantities. The elements of the matrix are the associated cost or times *etc.*

Step 2

Put a <u>zero in each row and column</u> *i.e.* find any rows that do not contain a zero, identify the smallest element in each of these rows then form a new matrix by subtracting the smallest elements from the other elements in their respective rows. Repeat this procedure, this time finding columns that do not contain a zero and subtracting the smallest elements from their respective columns.

Step 3

Find the <u>independent zeros</u> *i.e.* find a row that contains one (uncrossed out) zero only and circle that zero, cross out all other zeros in the column that contains the circled zero. Repeat these steps until all the rows have been examined. Repeat this procedure, this time finding columns that contain one (uncrossed out) zero only and crossing out the zeros in the appropriate rows. The steps above must be repeated until all zeros in the matrix are either circled or crossed out.

If :

- **There are *n* Independent Zeros**
  The optimal assignment has been found and the algorithm **terminates**. The optimal assignment can be found by superimposing the positions of the independent zeros onto the original matrix. The optimal cost or time *etc.* can be found by finding the sum of the circled costs or times *etc.*

- **There are > *n* Independent Zeros**
  The optimal assignment is **non-unique**. The optimal assignment can be found by choosing **any** *n* of the independent zeros and then proceeding as above.

- **There are < *n* Independent Zeros**
  The optimal assignment has yet to be found and the algorithm proceeds to **Step 4**.

Step 4

Find an <u>improved assignment</u>. To do this draw onto the matrix the **minimum** number of horizontal and vertical lines needed to pass through <u>**all**</u> zeros. Find the smallest element in the matrix that does <u>**not**</u> have a line passing through it then form a new matrix by :

- Subtracting this element from all other elements that do <u>**not**</u> have a line passing through them.

- Adding this element to all elements that have <u>**two**</u> lines passing through them.

Step 5

If each row and column contains a zero go back to **Step 3** otherwise go back to **Step 2**.

11.3 Other Assignment Problems

Unbalanced Problems
An unbalanced problem is one in which there more quantities in one set than in the other *e.g.* problems where there 5 jobs to be assigned to 8 machines or 7 employees to be assigned to 6 tasks *etc.* Problems of this kind can be solved using the algorithm described above. However, before proceeding to **Step 2**, the matrix must be made **square** by adding rows or columns of zero elements.

Unavailable Assignments
Unavailable assignments occur for a variety of reasons. For example, some employees may not be qualified to complete all tasks, some machines may not be capable of processing all jobs, *etc.* Problems of this kind can be solved using the algorithm described above. The unavailable assignments are simply ignored.

Maximisation Problems
On some occasions the elements of the matrix will be incomes or profits *etc.* and the optimal assignment will be the one that maximises the total value. To solve a problem of this kind :

- Form the data into a matrix as before.

- Find the largest element in the matrix *i.e.* the largest income or profit *etc.*

- Form a new matrix by subtracting each element **from** the **largest value**. The largest value will become zero and the resulting matrix will represent a minimisation problem. This problem is the dual of the maximisation problem.

- Solve the dual problem problem using the Hungarian algorithm.

- Extract the optimal assignment by superimposing the positions of the independent zeros in the optimal (minimisation) matrix onto the original (maximisation) matrix. The optimal cost or time *etc.* can be found by finding the sum of the circled costs or times in the original (maximisation) matrix.

Example 11.5
A company employs five salesmen to work in five regions of the UK. Due to differences in experience and ability it is not anticipated that each salesman will be equally effective in each area. The table below shows the estimated annual sales (in thousands of pounds) for each salesman in each region :

		Region				
		South-West	South-East	Midlands	North-West	North-East
Salesman	Doreen Ford	3	9	2	3	7
	Paul Burns	6	1	5	6	6
	Randeep Singh	9	4	7	10	3
	Lesley Stephens	2	5	4	2	1
	Gordon Murray	9	6	2	4	6

Find the assignment of salesmen to regions that will maximise that total sales figure.

Solution

The initial matrix for the maximisation problem is :

$$\begin{bmatrix} 3 & 9 & 2 & 3 & 7 \\ 6 & 1 & 5 & 6 & 6 \\ 9 & 4 & 7 & 10^* & 3 \\ 2 & 5 & 4 & 2 & 1 \\ 9 & 6 & 2 & 4 & 6 \end{bmatrix}$$

The largest element in this matrix is the 10 in row 3, column 4. Subtracting each element from this value, the initial matrix for the minimisation problem is :

$$\begin{bmatrix} 7 & 1 & 8 & 7 & 3 \\ 4 & 9 & 5 & 4 & 4 \\ 1 & 6 & 3 & 0 & 7 \\ 8 & 5 & 6 & 8 & 9 \\ 1 & 4 & 8 & 6 & 4 \end{bmatrix}$$

Solving the minimisation problem using the Hungarian algorithm the optimal matrix is :

$$\begin{bmatrix} 5 & (0) & 6 & 5 & 1 \\ \cancel{0} & 6 & 1 & \cancel{0} & (0) \\ 1 & 7 & 3 & (0) & 7 \\ 2 & \cancel{0} & (0) & 2 & 3 \\ (0) & 4 & 7 & 5 & 3 \end{bmatrix}$$

Superimposing the positions of the independent zeros onto the original (maximisation) matrix :

$$\begin{bmatrix} 3 & (9) & 2 & 3 & 7 \\ 6 & 1 & 5 & 6 & (6) \\ 9 & 4 & 7 & (10) & 3 \\ 2 & 5 & (4) & 2 & 1 \\ (9) & 6 & 2 & 4 & 6 \end{bmatrix}$$

The optimal assignment is :

Salesman	Region
Doreen Ford	South-East
Paul Burns	North-East
Randeep Singh	North-West
Lesley Stephens	Midlands
Gordon Murray	South-West

The total sales figure is 9+6+10+4+9 = 38 = £38,000

11.4 Exercises 11

1. In a small company there are five workers who must be assigned to five tasks. The estimated time required (in hours) by each worker to complete each task is shown in the table below :

		Task				
		A	B	C	D	E
Worker	1	3	8	2	10	3
	2	8	7	2	9	7
	3	6	4	2	7	5
	4	8	4	2	3	5
	5	9	10	6	9	10

Find the optimal assignment of workers to tasks and the total time taken.

2. A haulage company has five trucks I, II, III, IV and V for transporting sand to five building sites A, B, C, D and E. The cost (in £) of transporting a load of sand by each truck to each building site is shown in the table below :

		Building Site				
		A	B	C	D	E
Truck	I	100	60	40	80	30
	II	20	90	70	70	60
	III	60	110	120	50	90
	IV	50	40	20	10	40
	V	50	90	80	50	90

Find the optimal assignment of trucks to building sites and the total transport cost.

3. The radio dispatcher for Citicabs in St Albans has one taxi waiting at each of four locations A, B, C and D. Three customers at locations 1, 2, and 3 have telephoned the Citicabs office requesting service. The distance (in miles) between each waiting taxi and each customer is shown in the table below :

		Customer Location		
		1	2	3
Taxi Location	A	4	4	3
	B	2	5	5
	C	1	5	8
	D	4	3	7

Find the optimal assignment of taxis to customers and the total distance travelled.

4. Consider the problem of assigning four operators to four machines. The cost of assigning each operator to each machine (in £/hour) is shown in the table below :

		Machine			
		Grinder	Lathe	Milling Machine	Drill
Operator	Dave	5	5	-	2
	Sue	7	4	2	3
	Ben	9	3	5	-
	Sandeep	7	2	6	7

Dave and Ben are new employees and have yet to be trained to use the milling machine and the drill respectively. Find the optimal assignment of operators to machines and the total cost/hour.

5. A company in China manufactures a range of sports clothing that it sells in south-east Asia. The table below shows the profit (in $) made by the company by selling each item in each country :

		Country					
		Hong Kong	Taiwan	Vietnam	Cambodia	Malaysia	Indonesia
Garment	Hoodie	11	10	8	9	3	7
	Sweatshirt	3	9	8	6	5	10
	T-Shirt	6	8	6	7	7	5
	Shorts	5	10	6	20	12	9
	Cap	11	7	12	10	11	8
	Socks	7	5	4	10	11	6

Find the optimal sales strategy and the total profit.

6. A company employs five new members of staff who must be assigned to six empty offices (one member of staff per office). On their first day the staff are given a tour of the vacant offices and are asked to rank them on a simple scale *i.e.* 6 = first choice, 5 = second choice, *etc.* The results are shown in the table below :

		Office					
		201	207	211	212	219	231
Employee	Yvonne Percival	6	2	1	4	5	3
	James Kingsley	2	3	5	1	4	6
	Emily Deanus	6	3	5	2	1	4
	Graham Roberts	5	4	3	2	1	6
	Derek Johnson	4	5	3	6	1	2

Since the new employees are of equal status, the Head of Department decides to allocate the staff to the offices in the way that maximises the total of their preferences. Find the optimal assignment of employees to offices.

Part Three - Transportation Problems

12. Solving Transportation Problems

12.1 Introduction
A transportation problem involves a series of supply points (*e.g.* warehouses, factories, *etc.*) each containing a quantity of a (single) good and a series of demand points (*e.g.* shops, restaurants, *etc.*) each requiring a quantity of that good. There is a **route** between each supply point and demand point and each route has a **shipping cost** (per item) associated with it. The aim when solving a transportation problem is to find the **optimal shipping pattern** and the **total shipping cost**. Unless stated otherwise the optimal shipping pattern is the one that **minimises** the total shipping cost. To illustrate how problems of this kind are solved consider the following example.

The Bakery Problem
Three grocery shops, Brown's, Wallace's and Robinson's purchase bread in boxes from three bakeries, United, Allied and International. The supply, demand and shipping cost data (in £ per box) for this transportation problem are given below in Table 1.

		Shop (*i.e.* Demand)			
		Brown's	Wallace's	Robinson's	Total Supply
Bakery (*i.e.* Supply)	United	6	8	4	6
	Allied	4	9	3	10
	International	1	2	6	15
	Total Demand	14	12	5	

Table 1

Solving a Transportation Problem
A transportation problem can be formulated as a linear programming model. For example, suppose that :

- There are n supply points and m demand points.

- There is a supply s_i at each supply point, $i = 1, \ldots, n$.

- There is a demand d_j at each demand point, $j = 1, \ldots, m$.

- The amount transported from supply point i to demand point j is x_{ij}.

- The shipping cost for the route between supply point i and demand point j is c_{ij}.

- The total shipping cost is z.

Then, if **supply is equal to demand**, the optimal shipping pattern can be found by solving the following linear programming model :

$$\text{Minimise} \quad z = \sum_{i=1}^{n} \sum_{j=1}^{m} c_{ij} x_{ij}$$

$$\text{Subject to :} \quad \sum_{j=1}^{m} x_{ij} = s_i, \quad i = 1, \ldots, n \quad \textit{i.e. supplies must be exhausted.}$$

$$\sum_{i=1}^{n} x_{ij} = d_j, \quad j = 1, \ldots, m \quad \textit{i.e. demands must be met.}$$

$$x_{ij} \geq 0 \quad \forall i, j$$

This linear programming model is similar to the one associated with the assignment problem. In the same way it can be very large (even for small values of n and m) and is highly degenerate.

12.2 The Transportation Algorithm
To overcome these problems a special procedure has been developed called the **transportation algorithm**. Like the Hungarian algorithm described in Chapter 11 it takes advantage of the special structure and sparsity of a transportation problem, it can overcome degeneracy and is much more computationally efficient *i.e.*

- The transportation algorithm requires a tableau with n rows and m columns only. The corresponding linear programming model would require a tableau with $n+m$ rows and $n+m+nm$ columns.

- In the transportation algorithm no artificial variables are required and it is relatively easy to find an initial solution *i.e.* the transportation algorithm is a one-phase procedure.

- The transportation algorithm uses addition and subtraction only whereas the two-phase simplex method uses addition, subtraction, multiplication and division. Addition and subtraction can be performed much more accurately and quickly on a computer than multiplication and division.

Note
The transportation algorithm is also known as the **stepping stone algoithm**.

Steps in the Transportation Algorithm

Step 1 : Form the initial tableau
The rows of the tableau represent the supply of the good and the columns of the tableau represent the demand for the good. The shipping cost associated with each route is usually written in the top left-hand corner of the appropriate cell.

Example 12.1
The initial tableau for the bakery problem is :

	Demands			s_i
6	8	4		6
4	9	3		10
1	2	6		15
d_j	14	12	5	

Supplies

Table 2

Step 2 : Find an initial solution

The North-West Corner Method
To find an initial solution using the **North-West corner method** :

1) Look at the route in the North-West (*i.e.* top left-hand) corner of the tableau.

2) By looking at the supply and demand figures associated with this route, move the maximum amount of the good into it. It is usual to put a circle around the quantity entered. This figure is then called a **stepping stone**.

3) If the total supply for that row has been exhausted, cross out the remaining routes in that row. If the total demand for that column has been satisfied, cross out the remaining routes in that column.

4) Go back to step 1) and repeat the process on that part of the tableau that remains.

5) Repeat steps 1) to 5) above until all supplies have been exhausted and all demands have been satisfied.

Example 12.2
Using the North-West corner method the initial solution of the bakery problem is :

	Demands			s_i
6 — ⑥	8 — X	4 — X	6
4 — ⑧	9 — ②	3 — X	10
1 — X	2 — ⑩	6 — ⑤	15
d_j	14	12	5	

*(left label: **Supplies**)*

Table 3

Step 3 : Test for feasibility
Suppose that a transportation problem has n supply points and m demand points. Then, the current solution is **feasible** if and only if the tableau contains $n+m-1$ stepping stones. If the stepping stones extend from the North-West corner down to the South-East (*i.e.* bottom right-hand) corner of the tableau, the solution is called a **basic feasible solution**.

Example 12.3
In the bakery problem there are three supply points and three demand points. The initial solution produced using the North-West corner method contains $3+3-1$ *i.e.* 5 stepping stones. Hence, the initial solution is feasible. Furthermore, since the stepping stones extend from the North-West corner down to the South-East corner of the tableau, the initial solution is also a basic feasible solution.

Step 4 : Calculate the dual variables
To calculate the dual variables :

1) Add an additional row and column to the tableau for these variables. The row is usually labelled v and the column is usually labelled u.

2) Set the dual variable for the first row to zero.

3) Calculate the dual variables for each route **containing a stepping stone**. To do this assign u and v values so that $u+v$ is equal to the **shipping cost** for that route.

216

Example 12.4
Calculating the dual variables for the bakery problem :

	Demands			s_i	u
	6 ⟨6⟩	8	4	6	0
Supplies	4 ⟨8⟩	9 ⟨2⟩	3	10	-2
	1	2 ⟨10⟩	6 ⟨5⟩	15	-9
d_j	14	12	5		
v	6	11	15		

Table 4

Step 5 : Calculate the shadow costs

The **shadow costs** are the cost savings (per item) that can be achieved by using the routes on the tableau that do not currently contain stepping stones. These values are calculated by subtracting $u + v$ from the shipping cost associated with each available route. The shadow costs are written within the appropriate cells on the tableau but are **not** circled (to distinguish them from stepping stones).

Example 12.5
Calculating the shadow costs for the bakery problem :

	Demands			s_i	u
	6 ⟨6⟩	8 -3	4 -11	6	0
Supplies	4 ⟨8⟩	9 ⟨2⟩	3 -10	10	-2
Shadow Costs	1 4	2 ⟨10⟩	6 ⟨5⟩	15	-9
d_j	14	12	5		
v	6	11	15		

Table 5

217

Step 6 : Test for optimality
The current shipping pattern is optimal if **all** of the **shadow costs** in the tableau **are greater than or equal to zero**. If this condition is satisfied then **stop**. The optimal shipping pattern can be read off directly from the tableau and the total shipping cost can be calculated by multiplying the quantity inside each stepping stone by the shipping cost for that route and then adding these values together.

Example 12.6
Looking at the shadow costs in Table 5 it can be seen that the current shipping pattern for the bakery problem is **not** optimal.

Step 7 : Perform a rooke's tour
If the optimality condition is not satisfied, the solution must be improved by moving some of the good being transported to a cheaper route. The procedure used to do this is called a **rooke's tour** and is performed as follows :

1) Find the route that gives the largest cost saving. This is the route with the largest negative shadow cost.

2) Treating this route as if it has a stepping stone in it, look at the rows in the tableau and cross out any that do not contain at least **two** stepping stones. Repeat this operation for the columns in the tableau. The stepping stones that remain in the tableau after this operation are those that can be used to complete the transfer.

3) Put a "+" sign in the route chosen in step 1).This indicates that the quantity of the good to be transported on that route is to be increased.

4) Put a "+" sign or a "-" sign in the remaining stepping stones to indicate the adjustments necessary to rebalance the row and column totals. It may <u>not</u> be necessary to use <u>all</u> of the remaining stepping stones in this operation.

5) Find the smallest **decreasing** stepping stone and move its quantity into the route chosen in 1).

6) Adjust the remaining stepping stones to rebalance the row and column totals in the tableau. For some stepping stones this adjustment can be <u>zero</u>.

Example 12.7

Performing a rooke's tour on the tableau for the bakery problem :

No rows or columns are crossed out during the elimination process

Supplies

Adjustment necessary

Route that gives the largest cost saving

New stepping stone

Smallest decreasing stepping stone

	Demands			s_i	u
6 — ⑥ (−)	8 — −3	4 — −11 (+)		6	0
4 — ⑧ (+)	9 — ② (−)	3 — −10		10	−2
1 — 4	2 — ⑩ (+)	6 — ⑤ (−)		15	−9
d_j : 14	12	5			
v : 6	11	15			

Table 6

	Demands			s_i
6 — ④	8	4 — ②		6
4 — ⑩	9	3		10
1	2 — ⑫	6 — ③		15
d_j : 14	12	5		

Table 7

Step 8 : Go back to Step 4

Summary of the Transportation Algorithm

The transportation algorithm can be summarised as follows :

Step 1

Form the initial tableau. The rows of the tableau represent the supply of the good and the columns of the tableau represent the demand for the good. The shipping cost associated with each route is usually written in the top left-hand corner of the appropriate cell.

Step 2

Find an initial solution *e.g.* using the North-West corner method. Other methods for finding the initial solution will be considered later in this chapter.

Step 3

Test for feasibility. If the transportation problem has n supply points and m demand points, the current solution is feasible if and only if the tableau contains $n + m - 1$ stepping stones.

Step 4

Calculate the dual variables. Add an additional row and column to the tableau. The row is usually labelled v and the column is usually labelled u. Set the dual variable for the first row to zero. Then, for each route **containing** a stepping stone, assign the u and v values so that $u + v$ is equal to the shipping cost for that route.

Step 5

Calculate the shadow costs. Subtract $u + v$ from the shipping cost for each available route *i.e.* each route **without** a stepping stone. The shadow costs are written within the appropriate cells on the tableau but are **not** circled (to distinguish them from stepping stones).

Step 6

Test for optimality. The current shipping pattern is optimal if **all** of the **shadow costs** in the tableau are **greater than or equal to zero**. If this condition is satisfied then **stop**. The optimal shipping pattern can be read off directly from the tableau and the total shipping cost can be calculated by multiplying the quantity inside each stepping stone by the shipping cost for that route and then adding these values together.

Step 7

Improve the current solution by moving some of the good being transported to a cheaper route *i.e.* perform a **rooke's tour**. Find the route that gives the largest cost saving *i.e.* the route with the largest negative shadow cost. Treating this route as if it has a stepping stone in it, look at the rows in the tableau and cross out any that do not contain at least **two** stepping stones. Repeat this operation for the columns in the tableau. Put a "+" sign or a "-" sign in the remaining stepping stones to indicate the adjustments necessary to rebalance the row and column totals. It may not be necessary to use <u>all</u> of the remaining stepping stones in this operation. Find the smallest **decreasing** stepping stone and move its quantity into the route that gives the largest cost saving. Adjust the remaining stepping stones to rebalance the row and column totals.

Step 8

Go back to **Step 4**.

Note

The optimal solution of a transportation problem is **non-unique** if for one of the routes, the shadow cost is equal to the shipping cost.

Supplementary Exercise

Use Table 8 and Table 9 below to complete the solution of the bakery problem.

	Demands			s_i	u
	6	8	4		
				6	
	4	9	3		
Supplies				10	
	1	2	6		
				15	
d_j	14	12	5		
v					

Table 8

	Demands			s_i	u
	6	8	4		
				6	
	4	9	3		
Supplies				10	
	1	2	6		
				15	
d_j	14	12	5		
v					

Table 9

221

Solution

Table 8

Table 9

Since all of the shadow costs are positive the solution above is optimal. The total shipping cost $= £((1 \times 6) + (5 \times 4) + (10 \times 4) + (3 \times 1) + (12 \times 2)) = £93$.

12.3 Exercises 12a

1. Three factories in the West Midlands manufacture steel panels for the motor industry. The completed panels are stored in three warehouses until they are needed for production. The table below shows the production, storage and shipping cost data (in £ per panel) :

		Warehouse			
		1	2	3	Production
Factory	1	8	15	7	15
	2	10	6	10	30
	3	9	13	7	15
Storage Capacity		25	15	20	

Use the transportation algorithm to find the allocation of steel panels to warehouses that minimises the total shipping cost.

2. Three refineries supply petrol by road tanker to three supermarket service stations. The table below shows the daily availability of petrol (in 1000 litre units) at each refinery, the daily requirement for petrol (in 1000 litre units) at each service station and the shipping cost data (in £ per 1000 litres) :

		Service Station			
		Sainsbury's	ASDA	Tesco	Availability
Refinery	1	8	5	6	600
	2	10	4	8	400
	3	7	8	4	500
Requirement		500	250	750	

Use the transportation algorithm to find the allocation of petrol to service stations that minimises the total shipping cost.

3. In a small engineering company three grades of craftsmen are used for manufacturing three types of valves. The table below shows the number of craftsmen of each grade employed by the company number, the number of craftsmen required to manufacture each type of valve, and the cost (in £) of using each grade of craftsmen to manufacture each type of valve :

		Valve			
		Type 1	Type 2	Type 3	Availability
Grade	Junior	40	38	36	5
	Senior	42	36	37	10
	Principal	44	42	43	30
Craftsmen Required		10	15	20	

Use the transportation algorithm to find the allocation of craftsmen to valves that minimises the total manufacturing cost.

4. An electrical contractor has four jobs in progress on a large industrial estate. For security and administrative reasons he has arranged for his employees to be paid each week at three pay stations located around the estate. The table below shows the number of employees working on each job, the capacities of the pay stations and the average time taken (in minutes) for an employee to walk from each job to each pay station :

		Pay Station			
		1	2	3	Number Employed
Contract	A	10	6	3	166
	B	7	2	5	218
	C	4	13	7	149
	D	6	10	4	190
	Capacities	100	223	400	

The contractor wishes to minimise the total time lost when his employees collect their wages each week. Use the transportation algorithm to solve this problem.

12.4 Alternative Methods For Finding An Initial Solution

The Matrix Minimum Method
To find an initial solution using the **matrix minimum method** :

1) Find the route that has the lowest shipping cost. In the event of tie choose the route that has the largest capacity.

2) By looking at the supply and demand figures associated with this route, move the maximum amount of the good into it.

3) Look at the remaining routes in this row. Using the cheapest routes first, allocate the remaining supply for this row.

4) Cross out the remaining routes in this row to make them unavailable.

5) Cross out the remaining routes in any column whose demand has been satisfied.

6) Go back to step 1) and repeat the process on that part of the tableau that remains.

7) Repeat steps 1) to 6) above until all supplies have been exhausted and all demands have been satisfied.

Example 12.8

Using the matrix minimum method the initial solution of the bakery problem is :

	Demands			s_i
	6 / X	8 / (6)	4 / X	6
Supplies	4 / X	9 / (5)	3 / (5)	10
	1 / (14)	2 / (1)	6 / X	15
d_j	14	12	5	

Table 10

Vogel's Rule

Vogel's rule is the most efficient procedure for finding an initial solution of a transportation problem since the initial shipping pattern it produces is the nearer to the optimal solution than those produced by the North-West corner method and the matrix minimum method. To find an initial solution using this method :

1) For each row and column, subtract the smallest shipping cost from the next smallest shipping cost. These values are called the **penalty costs** and are the penalties associated with not using the cheapest route in that row or column.

2) Find the row or column that has the largest penalty cost and identify its cheapest route. If two rows or columns both have the largest penalty cost, choose one that contains the cheapest route. In the event of tie here, choose the route that has the largest capacity.

3) By looking at the supply and demand figures associated with this route, move the maximum amount of the good into it.

4) If the total supply for that row has been exhausted, cross out the remaining routes in that row. If the total demand for that column has been satisfied, cross out the remaining routes in that column.

5) Go back to step 1) and repeat the process on that part of the tableau that remains.

6) Repeat steps 1) to 5) above until all supplies have been exhausted and all demands have been satisfied.

Example 12.9
Using Vogel's rule the initial solution of the bakery problem is :

	Demands			s_i	Penalty Costs
	6 ⓵	8 X	4 ⑤	6	6-4=2 6-4=2 6-4=2
Supplies	4 ⑩	9 X	3 X	10	4-3=1 4-3=1 4-3=1
	1 ③	2 ⑫	6 X	15	2-1=1 6-1=5 -
d_j	14	12	5		
Penalty Costs	4-1=3 4-1=3 6-4=2	8-2=6 - -	4-3=1 4-3=1 4-3=1		

Table 11

Note
In this example Vogel's rule has produced the optimal solution. However, this is not normally the case.

12.5 Degeneracy
If a transportation problem has n supply points and m demand points the tableau must always contain exactly $n+m-1$ stepping stones. If at any point during the solution the tableau contains more (or less) than this number of stepping stones the solution is said to be **degenerate**. When a solution is degenerate **cycling** can occur and it can become impossible to calculate the dual variables.

Dealing With Degeneracy

Case 1 : When there are <u>more</u> than $n+m-1$ stepping stones.
In this case the tableau will contain a **loop** *e.g.*

	Demands			s_i	u
	1 [9]	**2** [1]	**4**	10	0
Supplies	**7**	**1** [5] ↔	**6** [10]	15	-1
	2	**4** [9] ↔	**3** [2]	11	**2 or -4 ?**
d_j	9	15	12		Loop
v	1	2	7		

Table 12

To overcome this problem stepping stones must be <u>removed</u> from the tableau to reduce the total number to $n+m-1$. This is done by moving goods around the loop until the required number of stepping stones become zero and can be removed. For example, to remove the degeneracy from the tableau above <u>one</u> stepping stone must be removed. To remove the stepping stone between supply point 3 and demand point 3 the quantity 2 must be moved around the loop as follows :

	Demands			s_i	u
	1 [9]	**2** [1]	**4**	10	0
Supplies	**7**	**1** [3] ↔	**6** [12]	15	-1
	2	**4** [11] ↔	**3** [0]	11	**2 or -4 ?**
d_j	9	15	12		
v	1	2	7		

Table 13

The zero stepping stone can now be removed and the dual variables can be calculated *i.e.*

	Demands			s_i	u
	1	**2**	**4**	10	0
	(9)	(1)			
Supplies	**7**	**1**	**6**	15	-1
		(3)	(12)		
	2	**4**	**3**	11	2
		(11)			
d_j	9	15	12		
v	1	2	7		

Table 14

The optimal shipping pattern can now be obtained in the usual way.

Case 2 : When there are <u>fewer</u> than $n+m-1$ stepping stones.
For example, consider the following tableau :

	Demands			s_i	u
	3	**3**	**4**	12	0
	(12)				
Supplies	**6**	**1**	**5**	9	?
		(8)	(1)		
	4	**2**	**2**	8	?
			(8)		
d_j	12	8	9		
v	3	?	?		

Table 15

To overcome this problem stepping stones must be <u>added</u> to the tableau to increase the total number to $n+m-1$. For each one this is done by finding the cheapest route that is not in use (and that will **not** produce a loop) and putting a zero stepping stone into it. For example, to remove the degeneracy from the tableau above <u>one</u> stepping stone must be added. The cheapest route that is not in use (and that will **not** produce a loop) is the route between supply point 1 and demand point 2. Putting a zero stepping stone into this route produces the following tableau :

Table 16

The dual variables can now be calculated *i.e.*

Table 17

The optimal shipping pattern can now be obtained in the usual way.

Supplementary Exercise
Investigate what happens if instead of the solution above, a zero stepping stone is put into the route between supply point 3 and demand point 2.

Answer
A loop is created in the tableau and it becomes impossible to calculate the dual variables.

Preventing Degeneracy

The Perturbation Technique
In the **perturbation technique** a small positive quantity ε is added to each supply s_i. To maintain equality between supply and demand the quantity $n\varepsilon$ (where n is the number of supply points) is added to the last demand d_m. The transportation algorithm is then applied to the perturbed problem in the usual way. When the optimal shipping pattern has been found the quantity ε is assigned the value **zero**.

Note
Adding the quantity ε to each supply s_i prevents degeneracy by ensuring that zeros do not appear on the right-hand sides of the constraints in the linear programming formulation of a transportation problem. This in turn, ensures that the transportation tableau always contains exactly $n+m-1$ stepping stones.

Example 12.10
Use the perturbation technique to solve the following transportation problem. Find the initial solution using the North-West corner method.

Supplies	Demands				s_i
	5	4	3	2	5
	10	8	4	7	5
	9	9	8	4	5
d_j	1	4	4	6	

Table 18

Solution

The perturbed problem is :

Table 19

Using the North-West corner method the initial solution is :

Table 20

Applying the transportation algorithm :

	Demands			s_i	u
5 — ①1	4 — ④4	3 — ⓔ ε (−)	2 — −4 — (dotted ◯) (+)	$5+\varepsilon$	0
10	8 [4]	4 [3] — (4−ε) (+)	7 — (1+2ε) (−)	$5+\varepsilon$	1
9	9 [6]	8 [7]	4 — (5+ε)	$5+\varepsilon$	−2
d_j : 1	4	4	$1+2\varepsilon$		
v : 5	4	3	6		

Table 21

	Demands			s_i
5 — ①1	4 — ④4	3	2 — ⓔ ε	$5+\varepsilon$
10	8	4 — ④4	7 — (1+ε)	$5+\varepsilon$
9	9	8	4 — (5+ε)	$5+\varepsilon$
d_j : 1	4	4	$6+3\varepsilon$	

Table 22

Table 23

	Demands				s_i	u
Supplies	[5] (1)	[4] (4) −	[3] 4	[2] (ε) +	$5+\varepsilon$	0
	[10] 0	[8] −1 (⋯) +	[4] (4)	[7] (1+ε) −	$5+\varepsilon$	5
	[9] 2	[9] 3	[8] 7	[4] (5+ε)	$5+\varepsilon$	2
d_j	1	4	4	$6+3\varepsilon$		
v	5	4	−1	2		

Table 23

Table 24

	Demands				s_i	u
Supplies	[5] (1)	[4] (3−ε)	[3]	[2] (1+2ε)	$5+\varepsilon$	
	[10]	[8] (1+ε)	[4] (1+ε)	[7]	$5+\varepsilon$	
	[9]	[9]	[8]	[4] (5+ε)	$5+\varepsilon$	
d_j	1	4	4	$6+3\varepsilon$		

Table 24

233

	Demands				s_i	u
5 [1]	4 [$3-\varepsilon$]	3	3 2 [$1+2\varepsilon$]		$5+\varepsilon$	0
10 1	8 [$1+\varepsilon$]	4 [$1+\varepsilon$]	7 1		$5+\varepsilon$	4
9 2	9 3	8 6	4 [$5+\varepsilon$]		$5+\varepsilon$	2
d_j 1	4	4	$6+3\varepsilon$			
v 5	4	0	2			

(Supplies label on left)

Table 25

Since all of the shadow costs are positive the solution above is optimal. Let $\varepsilon = 0$. Then the optimal shipping pattern is :

	Demands				s_i
5 [1]	4 [3]	3	3 2 [1]		5
10 1	8 [1]	4 [4]	7 1		5
9 2	9 3	8 6	4 [5]		5
d_j 1	4	4	6		

(Supplies label on left)

Table 26

The total shipping cost $= \pounds((1\times5)+(3\times4)+(1\times2)+(1\times8)+(4\times4)+(5\times4)) = \pounds63$.

12.6 Variation Between Supply and Demand

In the bakery problem the total supply of the good is the same as the total demand. In many transportation problems this is not the case *i.e.* supply will sometimes exceed demand and vice-versa. To overcome this problem the tableau must be extended to compensate for the difference between these values *i.e.*

If Supply > Demand

An additional **column** is added to the tableau to represent a **fictitious destination** whose demand matches the surplus. The shipping costs in the new column are set to the **storage costs** associated with the excess. However, in the absence of this information the shipping costs in the new column are set to zero. The optimal shipping pattern is then found in the usual way (avoiding using the fictitious routes if possible). The stepping stones in the additional column give the surplus at each supply point.

Example 12.11

Suppose that the supply and demand figures in the bakery problem are changed as follows :

Table 27

In this case the total supply exceeds the total demand by $35 - 31$ *i.e.* 4 boxes. To overcome this problem a fictitious destination is added to the tableau whose demand matches the surplus *i.e.*

Table 28

Fictitious Destination

The total supply of the good is now the same as the total demand.

Supplementary Exercise
Use the transportation algorithm to find the optimal shipping pattern in this case. Find the initial solution using Vogel's rule. Do **not** use the zero shipping costs when calculating the penalty costs as this encourages the use of the fictitious routes.

Solution
The optimal shipping pattern is :

	Demands				s_i
	6 (1)	8	4 (5)	0 (4)	10
Supplies	4 (10)	9	3	0	10
	1 (3)	2 (12)	6	0	15
d_j	14	12	5	4	

Table 29

The total shipping cost is £93.

If Demand > Supply
An additional **row** is added to the tableau to represent a **fictitious source** whose supply matches the shortfall. The shipping costs in the new row are set to the **penalty costs** associated with the deficit. However, in the absence of this information the shipping costs in the new row are set to zero. The optimal shipping pattern is then found in the usual way (avoiding using the fictitious routes if possible). The stepping stones in the additional row give the shortfall at each demand point.

Example 12.12

Suppose that the supply and demand figures in the bakery problem are changed as follows :

	Demands			s_i
Supplies	6	8	4	6
	4	9	3	10
	1	2	6	15
d_j	14	12	10	

Table 30

In this case the total demand exceeds the total supply by $36-31$ *i.e.* 5 boxes. To overcome this problem a fictitious source is added to the tableau whose supply matches the deficit *i.e.*

	Demands			s_i
Supplies	6	8	4	6
	4	9	3	10
	1	2	6	15
	0	0	0	5 ◄── Fictitious Source
d_j	14	12	10	

Table 31

The total supply of the good is now the same as the total demand.

Supplementary Exercise

Use the transportation algorithm to find the optimal shipping pattern in this case. Find the initial solution using the matrix minimum method.

Solution

The optimal shipping pattern is :

Demands			s_i
6	8	4 ⟨6⟩	6
4 ⟨6⟩	9	3 ⟨4⟩	10
1 ⟨8⟩	2 ⟨7⟩	6	15
0	0 ⟨5⟩	0	5
d_j 14	12	10	

Table 32

The total shipping cost is £82.

12.7 Doig's Paradox

Consider the following transportation problem :

Demands		s_i
300	60	8
50	320	7
d_j 5	10	

Table 33

The optimal shipping pattern for this problem is :

238

Table 34

	Demands		s_i
Supplies	300	60 (8)	8
	50 (5)	320 (2)	7
d_j	5	10	

The total shipping cost is £1370.

Suppose that the amount of goods being transported is **increased** so that the problem becomes :

Table 35

	Demands		s_i
Supplies	300	60	10
	50	320	7
d_j	7	10	

The optimal shipping pattern in this case is :

Table 36

	Demands		s_i
Supplies	300	60 (10)	10
	50 (7)	320	7
d_j	7	10	

The total shipping cost is £950.

It can be seen that **increasing** the amount of goods being transported has **reduced** the total shipping cost. This situation is known as **Doig's paradox** and occurs when increasing the supplies allows the demands to be met using fewer (and cheaper) routes.

Once the optimal solution of a transportation problem has been found it is relatively easy to determine whether Doig's paradox exists *i.e.*

Doig's paradox exists when in an optimal tableau, one or more of the $u+v$ values for the **unused** routes is **negative**.

For example, in the optimal tableau shown in Table 34 the $u+v$ value of the unused route between supply point 1 and demand 1 is -210 *i.e.*

	Demands		s_i	u
	300	60		
$u+v=-210$ →		⑧	8	0
Supplies				
	50	320		
	⑤	②	7	260
d_j	7	10		
v	-210	60		

Table 37

Hence, Doig's paradox exists within this problem.

When Doig's paradox is detected it is usual to identify the supply and demand totals that can be increased, the amount of the increase and the resulting cost saving. To do this :

1) Find the unused route that that has the largest negative $u+v$ value. The position of this route indicates the supply and demand totals that can be increased.

2) Find the transfer loop associated with the unused route using the procedure used in the rooke's tour *i.e.* treating this route as if it has a stepping stone in it, look at the rows in the tableau and cross out any that do not contain at least **two** stepping stones. Repeat this operation for the columns in the tableau.

3) Put a "-" sign in the unused route. Then put a "+" sign or a "-" sign in the remaining stepping stones to indicate the adjustments necessary to rebalance the row and column totals.

4) Find the smallest **decreasing** stepping stone and add its quantity to the supply and demand totals identified in 1).

5) Ignoring the unused route, adjust the remaining stepping stones on the tableau by the quantity identified in 4). This operation produces the new optimal shipping pattern.

6) Calculate the new total shipping cost and the resulting cost saving.

Applying this procedure to the optimal tableau shown in Table 37 :

Demands s_i u

Largest negative $u+v$ value   ⟶

	300	60		
	-210	(8)	8 ←	0
Supplies	50	320		
	(5)	(2)	7	260
d_j	7	10		
v	-210	60		

Supply and demand figures that can be increased

Table 38

Demands s_i

	300	60	
	( ) ⟷ (8)		8
	− +		
Supplies	50	320	
	(5) ⟷ (2)		7
	+ −		
d_j	7	10	

— Transfer loop

— Smallest decreasing stepping stone

— Supply and demand adjustment

Table 39

Hence, the totals for supply point 1 and demand point 1 can be increased by 2. Adjusting the stepping stones by this amount the new optimal tableau is :

Demands s_i

	300	60	
		(10)	8+2=10
Supplies	50	320	
	(7)		7
d_j	5+2=7	10	

Table 40

241

The new total shipping cost is $£((10\times60)+(7\times50))$ *i.e.* £950 and the resulting cost saving is $£1370-£950=£420$.

12.8 Other Transportation Problems

Maximisation Problems
In some transportation problems the shipping costs are incomes or profits *etc.* The optimal solution of a problem of this kind is the one that **maximises** the total cost. To solve a problem of this kind :

1) Find the route that has the largest shipping cost *i.e.* the largest income or profit *etc.*

2) Form an equivalent minimisation problem.

- The supply and demand values in this problem are the **same** as those in the maximisation problem.

- The shipping costs in this problem are calculated by subtracting each shipping cost in the maximisation problem **from** the largest shipping cost found in 1).

This problem is the dual of the maximisation problem.

3) Solve the dual problem using the transportation algorithm.

4) The optimal **shipping pattern** can then be extracted from the optimal **minimisation** tableau. The total **shipping cost** can be calculated by superimposing the optimal shipping pattern onto the initial **maximisation** tableau.

Example 12.13
A haulage company transports groceries from three depots to four supermarkets. The supply, demand and profit data (in £ per load) for this transportation problem are given in Table 41 below.

Depot		Supermarket 1	2	3	4	Supply
	1	6	4	5	4	6
	2	4	7	5	3	8
	3	2	6	5	2	4
Demand		4	3	5	6	

Table 41

Find the shipping pattern that maximises the total profit made by the company and the total profit.

242

Solution

The initial tableau for the maximisation problem is :

	Demands				s_i
	6	4	5	4	6
Supplies	4	7	5	3	8
	2	6	5	2	4
d_j	4	3	5	6	

Largest Profit →

Table 42

The largest profit is the £7 per load between depot 2 and supermarket 2. Subtracting the other profits from this value, the initial tableau for the minimisation problem is :

	Demands				s_i
	1	3	2	3	6
Supplies	3	0	2	4	8
	5	1	2	5	4
d_j	4	3	5	6	

Table 43

Solving this problem using the transportation algorithm, the optimal tableau is :

Demands

	1	2	3	4	s_i
Supplies	1 ⟨4⟩	3	2	3 ⟨2⟩	6
	3	0 ⟨3⟩	2 ⟨1⟩	4 ⟨4⟩	8
	5	1	2 ⟨4⟩	5	4
d_j	4	3	5	6	

Table 44

Superimposing the optimal shipping pattern onto the initial maximisation tableau :

Demands

	1	2	3	4	s_i
Supplies	6 ⟨4⟩	4	5	4 ⟨2⟩	6
	4	7 ⟨3⟩	5 ⟨1⟩	3 ⟨4⟩	8
	2	6	5 ⟨4⟩	2	4
d_j	4	3	5	6	

Table 45

Hence, the total profit made by the company is :

$$\pounds((4x6)+(2x4)+(3x7)+(1x5)+(4x3)+(4x5)) = \pounds90$$

Unavailable Routes
In the transportation problems described so far in this chapter the item being transported can be shipped between all supply points and all demand points without restriction. However, in some transportation problems certain routes are (or can become) unavailable. These routes must be ignored by the transportation algorithm. In a minimisation problem this is done by assigning large shipping costs to the unavailable routes (*e.g.* ∞). In a maximisation problem this is done by assigning large **negative** shipping costs to the unavailable routes (*e.g.* $-\infty$).

Transshipment Problems

A transshipment problem is a transportation problem in which the goods being shipped can be transported to their final destination via another source or destination. To solve a problem of this kind :

1) Find the shipping costs for all possible routes.

2) Extend the tableau so that the sources become possible destinations with zero demand and the destinations become possible sources with zero supply.

3) Enter the shipping costs. The shipping costs on the leading diagonal will be zero (since the cost of transporting goods from each location to itself is zero).

4) Calculate the total amount of goods being transported and add this value to each supply and demand figure on the tableau. This ensures that there are sufficient supplies available at each source should the optimal shipping pattern require that all goods originate from one source and/or be delivered to one destination.

5) Solve the extended problem using the transportation algorithm.

6) Extract the optimal shipping pattern. This is given in the part of the tableau that remains after the routes on the leading diagonal have been removed. The total shipping cost is then calculated in the usual way.

Example 12.14

Johnson's supermarket transports boxes of apples from warehouses in Glasgow and Leeds to distribution centres in Stoke, Norwich and Cardiff. The apples can be transported to their final destination via the other distribution centres if necessary. Adequate storage space is available for this purpose Table 46 shows the number of boxes available at each warehouse, the number of boxes required at each distribution centre and the shipping costs for all possible routes (in £ per box) :

	Glasgow	Leeds	Stoke	Norwich	Cardiff	Number
Glasgow	-	80	10	20	30	100
Leeds	10	-	20	30	40	200
Stoke	20	30	-	40	10	
Norwich	40	20	10	-	20	
Cardiff	60	70	80	20	-	
Number			100	100	100	

Table 46

The company manager wishes to find the transport plan that minimises the total shipping cost. Use the transportation algorithm to solve this transshipment problem. Use Vogel's rule to find the initial solution.

Solution

The total amount of goods being transported is $100+200=300$. Adding this value to each supply and demand figure the transshipment tableau becomes :

Demands

	s_1	s_2	d_1	d_2	d_3	s_i
s_1	0	80	10	20	30	400
s_2	10	0	20	30	40	500
Supplies d_1	20	30	0	40	10	300
d_2	40	20	10	0	20	300
d_3	60	70	80	20	0	300
d_j	300	300	400	400	400	

Table 47

Using Vogel's rule the initial solution is :

Demands

		s_1	s_2	d_1	d_2	d_3	s_i	Penalty Costs
	s_1	0 / 300	80 / X	10 / 100	20 / X	30 / X	400	10-0=10 10-0=10 20-10=10 10-0=10 10-0=10 10-0=10 10-0=10
	s_2	10 / X	0 / 300	20 / 100	30 / 100	40 / X	500	10-0=10 20-10=10 30-20=10 10-0=10 20-10=10 20-10=10 20-10=10
Supplies	d_1	20 / X	30 / X	0 / 200	40 / X	10 / 100	300	10-0=10 10-0=10 - 10-0=10 20-0=20 10-0=10 -
	d_2	40 / X	20 / X	10 / X	0 / 300	20 / X	300	10-0=10 - - 10-0=10 - 10-0=10 -
	d_3	60 / X	70 / X	80 / X	20 / X	0 / 300	300	20-0=20 - - - - - -
	d_j	300	300	400	400	400		
Penalty Costs		10-0=10 10-0=10 10-0=10 10-0=10 10-0=10 10-0=10 -	20-0=20 20-0=20 - - - - -	10-0=10 10-0=10 10-0=10 10-0=10 10-0=10 20-10=10 20-10=10	20-0=20 20-0=20 20-0=20 30-20=10 30-20=10 30-20=10 30-20=10	10-0=10 20-10=10 20-10=10 30-10=20 - - -		

Table 48

Calculating the dual variables and the shadow costs :

Demands

Supplies	s_1	s_2	d_1	d_2	d_3	s_i	u
s_1	0 (300)	80 90	10 (100)	20 0	30 10	400	0
s_2	10 0	0 (300)	20 (100)	30 (100)	40 10	500	10
d_1	20 30	30 50	0 (200)	40 30	10 (100)	300	-10
d_2	40 60	20 50	10 20	0 (300)	20 20	300	-20
d_3	60 80	70 100	80 90	20 20	0 0 (300)	300	-20
d_j	300	300	400	400	400		
v	0	-10	10	20	20		

Table 49

Since all of the shadow costs are greater than or equal to zero, this solution is optimal.

Ignoring the routes on the leading diagonal, the optimal shipping pattern is :

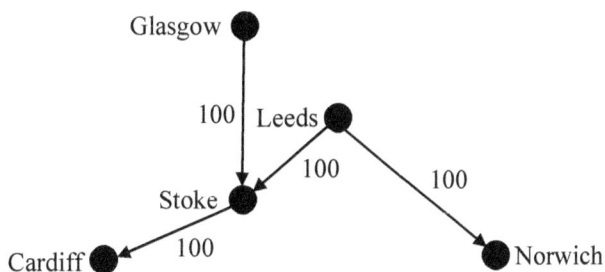

The total shipping cost is :

$$£((100\times10)+(100\times20)+(100\times30)+(100\times10)) = £7000$$

Capacitated Problems

A capacitated problem is one in which there is a limit on the amount of goods that can be transported along some or all of the routes. Problems of this kind are solved using the transportation algorithm **except that** :

- The capacitated routes are loaded to the maximum permitted values only.

- When a capacitated route is fully loaded :

  - A bar is put above the quantity written inside the stepping stone.

  - It is **ignored** when :

    - Checking for feasibility (*i.e.* that there are $n+m-1$ stepping stones).

    - Calculating the dual variables u and v.

    - Calculating the shadow costs.

    - Performing a rooke's tour (unless the quantity being transported on that route is being reduced).

Notes

- Capacitated problems are usually highly degenerate.

- The total shipping cost in a minimisation (maximisation) capacitated problem is usually greater (lower) than the total shipping cost in the corresponding uncapacitated problem.

Example 12.15

Consider the following transportation problem :

Table 50

The shipping costs here are in £. Find the optimal shipping pattern and the total shipping cost if the amount that can be transported between supply point 1 and demand point 1 is limited to 50 items.

Solution
Using the North-West corner method, the initial solution is :

North-West corner after the capacitated route is loaded

Fully loaded capacitated route →

	Demands			s_i
Supplies	5 — 50	7 — 100	4 — X	150
	4 — 100	2 — 50	5 — 50	200
	5 — X	6 — X	4 — 50	50
d_j	150	150	100	

Table 51

Ignoring the fully loaded capacitated route, this solution contains $n + m - 1 = 5$ stepping stones *i.e.* this solution is feasible.

Applying the transportation algorithm :

Route ignored →

	Demands			s_i	u
Supplies	5 — 50	7 — 100 (−)	4 4 — (−6) (+)	150	0
	4 — 100	2 — 50 (+)	5 — 50 (−)	200	−5
	5 2	6 5	4 — 50	50	−6
d_j	150	150	100		
v	9	7	10		

This stepping stone is not needed in the rooke's tour

Table 52

Table 53

	Demands			s_i	u
Route ignored →	5 — $\overline{50}$	7 — 50 (−)	4 — 50 (+)	150	0
Supplies	4 — 100 (−)	2 — 100 (+)	5 — 6 (−)	200	−5
	5 — −4 ( ) (+)	6 — −1	4 — 50 (−)	50	0
d_j	150	150	100		
v	9	7	4		

Table 54

	Demands			s_i	u
Route ignored →	5 — $\overline{50}$	7 — 4	4 — 100	150	0
Supplies	4 — 50	2 — 150	5 — 2	200	−1
	5 — 50	6 — 3	4 — 0	50	0
d_j	150	150	100		
v	5	3	4		

Zero stepping stone left on the tableau to remove degeneracy

Since all of the shadow costs are greater than or equal to zero, this solution is optimal.

The total shipping cost is :

$$£((50\text{x}5)+(100\text{x}4)+(50\text{x}4)+(150\text{x}2)+(50\text{x}5)) = £1400$$

Supplementary Exercise
Solve the transportation problem above ignoring the limit on the amount of goods that can be transported between supply point 1 and demand point 1. Compare the optimal shipping pattern and the total shipping cost obtained with those shown above.

Solution

The optimal shipping pattern is :

	Demands			s_i
	5 100	7	4 50	150
Supplies	4 50	2 150	5	200
	5	6	4 50	50
d_j	150	150	100	

Table 55

The total shipping cost is £1400. In this case the total shipping cost is the same as it was in the capacitated problem.

12.9 Exercises 12b

1. Find an initial solution of the transportation problem in question 1. of Exercises 12a using the matrix minimum method.

2. Find an initial solution of the transportation problem in question 2. of Exercises 12a using Vogel's rule. Is the initial solution optimal ? If so, calculate the total shipping cost.

3. Three secondary schools in Stevenage offer places to pupils from two local areas, Bedwell and Chells. The table below shows the number of places available at each school, number of children needing places in each area and the average time taken (in minutes) for a child to walk from each area to each school :

		Area		
		Bedwell	**Chells**	**Places Available**
School	**Barnwell**	5	7	150
	Shephalbury	9	6	250
	Heathcote	7	12	250
	Children in Area	400	250	

The Local Education Authority wishes to find the allocation of children to schools that will minimise the total time taken by the children to walk to school. Use the transportation algorithm to solve this problem. Find the initial solution using the North-West corner method.

4. Solve the transportation problem from question 3. using the perturbation technique. Find the initial solution using the North-West corner method.

5. Write down the linear programming formulation of the transportation problem in the case where :

(a) Supply > Demand.

(b) Demand > Supply.

6. British Rail transport coal from three mines in Yorkshire to four power stations in Scotland. The table below shows the number of coal wagons available at each mine, the number of coal wagons required by each power station and the shipping costs (in £ per wagon) :

		Power Station				
		Glasgow	Falkirk	Paisley	Aberdeen	Availability
Mine	Barnsley	10	23	14	13	14
	Harrogate	16	11	15	8	18
	Doncaster	17	18	9	12	32
	Requirement	15	17	26	12	

The transport manager at British Rail wishes to find transport plan that minimises the total shipping cost. Use the transportation algorithm to solve this problem. Find the initial solution using Vogel's rule. Use the perturbation technique to avoid degeneracy.

7. Two small factories in St Albans manufacture plastic tables and chairs for exterior use. The completed furniture is stored in two warehouses until it is collected for sale by local shops and garden centres. The table below shows the production, storage and shipping cost data (in pence per set) :

		Warehouse		
		1	2	Production
Factory	1	400	10	20
	2	20	500	5
	Storage Capacity	10	15	

(a) Use the transportation algorithm to find the allocation of furniture to warehouses that minimises the total shipping cost. Find the initial solution using the North-West corner method.

(b) Show that Doig's paradox exists within this problem.

(c) Identify the factory and warehouse totals that can be increased, the amount of the increase and the resulting cost saving.

8. A haulage company transports boxes of fruit from three warehouses in London to four supermarkets on the south coast. The table below shows the number of boxes available at each warehouse, the number of boxes required by each supermarket and the profits made by the haulage company (in pence per box) :

		Supermarket				
		Bournemouth	Brighton	Portsmouth	Southampton	Availability
Warehouse	Acton	6	6	6	4	1000
	Fulham	4	2	4	5	700
	Wapping	5	6	7	8	900
	Requirement	900	800	500	400	

The manager of the haulage company wishes to find the transport plan that maximises the total profit. Use the transportation problem to solve this problem. Find the initial solution using the North-West corner method.

9. Consider the transportation problem in question 1. of Exercises 12a. Suppose that the route between factory 1 and warehouse 1 is unavailable. Resolve the problem to find the optimal shipping pattern and the total shipping cost in this case.

10. The Ford motor company in the United States transports new cars from its factories in Los Angeles, Detroit and New Orleans to its distribution centres in Denver and Miami. Cars can be transported from the car plants to the distribution centres via the other car plants if necessary. Adequate storage space is available for this purpose The table below shows the number of cars available at each plant, the number of cars required at each distribution centre and the shipping costs for all possible routes (in $ per car) :

	Los Angeles	Detroit	New Orleans	Denver	Miami	Number
Los Angeles	-	130	90	80	215	1000
Detroit	135	-	101	100	108	1500
New Orleans	95	105	-	102	68	1200
Denver	79	99	110	-	205	
Miami	200	107	72	205	-	
Number				2300	1400	

The company manager wishes to find the transport plan that minimises the total shipping cost. Use the transportation algorithm to solve this transshipment problem. Find the initial solution using the North-West corner method.

11. Consider the transportation problem in question 1. of Exercises 12a. Suppose that the maximum number of steel panels that can be transported between factory 1 and warehouse 1 is 5. Resolve the problem to find the optimal shipping pattern and the total shipping cost in this case.

Answers to Exercises

Exercises 1a

1. (i) $x_1 = 12,\ x_2 = 0,\ z = 24$

    (ii) $x_1 = 2,\ x_2 = 4,\ z = 16$

    (iii) $x_1 = 1,\ x_2 = 2,\ z = 8$

2. (i) x_1 is the amount of Standard cloth, x_2 is the amount of Deluxe cloth.

    (ii)   The objective function gives the weekly profit made by the company. The constraints show the requirements on (and the availabilities of) the different wools. The non-negativity conditions show the factory cannot make a negative amount of each cloth.

    (iii) The company should make $480kg$ of the Standard cloth each week and $420kg$ of the Deluxe cloth each week. The maximum weekly profit is then £1110.

3. (i)   Let z be the total wage bill, x_1 be the number of days Robyn is employed and x_2 be the number of days Laura is employed. Then, the linear programming model is :

    Minimise $z = 25x_1 + 22x_2$

    Subject to : $x_1 + x_2 \geq 5$

    $\qquad\qquad\ \ 3x_1 + 2x_2 \geq 12$

    $\qquad\qquad\ \ 3x_1 + 6x_2 \geq 18$

    $\qquad\qquad\ \ x_1, x_2 \geq 0$

    (ii)   The manager should employ Robyn for 2 days and Laura for 3 days. The minimum total wage bill is then £116.

4. (i)   Let z be the total weekly profit, x_1 be the number of Majestic quilts made and x_2 be the number of Royal quilts made. Then, the linear programming model is :

    Maximise $z = 12x_1 + 10x_2$

    Subject to : $4x_1 + 2x_2 \leq 2000$

    $\qquad\qquad\ \ 3x_1 + 5x_2 \leq 3000$

    $\qquad\qquad\ \ x_1 \geq 100$

    $\qquad\qquad\ \ x_2 \geq 200$

    $\qquad\qquad\ \ x_1, x_2 \geq 0$

    (ii)   The company should manufacture 286 Majestic and 428 Royal each week. The maximum total weekly profit is then £7712.

    (iii) The profit made on each Majestic can vary from £6 to £20 inclusive.

(iv) The company should manufacture 357 Majestic and 385 Royal each week. The maximum total weekly profit is then £8134.

(v) The company would be prepared to pay (up to) the additional profit as rent
   i.e. £8134-£7712 = £422 each week.

Note : In 4(ii) and 4(iv) above the number of Royal quilts has been rounded down to ensure that the solution is feasible *i.e.* satisfies the constraints. Royal quilts are chosen for rounding because the company makes a smaller profit on these than they do on Majestic quilts *i.e.* rounding down the number of Royal quilts produces the smallest reduction in the total weekly profit. Rounding non-integer solutions of linear programming problems can be problematic. Alternative procedures for solving integer constrained problems are discussed in Chapter 8.

5. (i) All points on the line joining the points $(0,3)$ and $(1,5)$ *i.e.* the problem has infinitely many solutions.

(ii) The problem is infeasible.

(iii) The problem is unbounded.

Exercises 1b

1. (i) A basic solution is one that is obtained by setting some of the variables to zero and then solving the constraints *i.e.* the system of linear equations, for the remaining variables.

(ii) A non-basic variable is a variable whose value is set to zero.

(iii) A basic variable is a variable for which the system is solved.

(iv) A basic feasible solution is a basic solution that satisfies the non-negativity conditions.

(v) A slack variable is a variable that is added to a type 1 constraint to convert it into an equation. If the constraint represents the limit on the availability of a resource then the slack variable represents the unused amount of that resource.

(vi) A linear programming problem is said to be in canonical form if it is written as a minimisation problem in which each constraint is an equation that contains a basic variable with a coefficient of one.

2. Minimise $\quad -\bar{z} - 4x_1 + 3x_2 - 7x_3 - 2x_4 + x_5 = -125$
   Subject to : $\quad x_1 - x_3 + x_4 - 7x_5 + x_6 = 6$
   $$-6x_1 + x_2 - x_4 + 4x_5 + x_7 = 7$$
   $$5x_1 + 2x_3 - 7x_4 + x_8 = 8$$
   $$x_2 - 8x_3 + x_4 - 6x_5 + x_9 = 7$$
   $$x_1, x_2, x_3, x_4, x_5, x_6, x_7, x_8, x_9 \geq 0$$

3. (a) $z = 130, \; x_1 = 10, \; x_2 = 30$

4. (i) Let z be the weekly profit, x_1 be the amount of Breakfast Blend produced (in kg) and x_2 be the amount of After Dinner produced (in kg). Then, the linear programming model is :

Maximise $z = 25x_1 + 50x_2$
Subject to : $1/7 x_1 + 2/5 x_2 \leq 120$
$3/7 x_1 + 2/5 x_2 \leq 180$
$3/7 x_1 + 1/5 x_2 \leq 150$
$x_1, x_2 \geq 0$

(ii) Fortesque's should produce $210kg$ of Breakfast Blend each week and $225kg$ of After Dinner each week. The maximum weekly profit is then £165.

5. (i) Let z be the weekly profit, x_1 be the amount of Domestic rope produced (in m) and x_2 be the amount of Heavy Duty rope produced (in m). Then, the linear programming model is :

Maximise $z = 0.4x_1 + 0.25x_2$
Subject to : $3x_1 + 6x_2 \leq 1100$
$4x_1 + 7x_2 \leq 1900$
$5x_1 + 4x_2 \leq 1400$
$x_1, x_2 \geq 0$

(ii) The company should produce $280m$ of Domestic rope and $0m$ of Heavy Duty rope each week. The maximum weekly profit is then £112.

(iii) The unused amounts of nylon are : Grade 1 : $260g$, Grade 2 : $780g$, Grade 3 : $0g$.

6. (i) Let z be the daily profit, x_1 be the number of loads of Light purchased each day, x_2 be the number of loads of Light pre-processed into Heavy and x_3 be the number of loads of Heavy purchased each day. Then, the linear programming model is :

Maximise $z = 45x_1 + 30x_2 + 55x_3$
Subject to : $x_1 + x_3 \leq 30$
$x_1 \leq 20$
$x_2 + x_3 \leq 25$
$x_1, x_2, x_3 \geq 0$

Note : The constraint $x_3 \leq 30$ is redundant.

(ii)   The oil company should :

- Buy 20 loads of Light each day.

- Buy 10 loads of Heavy each day.

- Process 15 loads of Light into Heavy.

- Refine 5 loads of Light into Petrol and sell it.

- Refine 25 loads of Heavy into Diesel and sell it.

The maximum daily profit is then £1900.

Exercises 2

1. (i) $x_1 = 0$, $x_2 = 1$, $x_3 = 0$, $z = -5$

   (ii) $x_1 = 45$, $x_2 = 0$, $z = 45$

   (iii) $x_1 = 5$, $x_2 = 0$, $x_3 = 5$, $z = 3$

   (iv)   The problem is infeasible.

2. (i)   Let z be the total profit, x_1 be the number of Amber chairs manufactured, x_2 be the number of Beaton chairs manufactured and x_3 be the number of Countess chairs manufactured. Then, the linear programming model is :

Maximise $z = 10x_1 + 14x_2 + 11x_3$
Subject to : $0.3x_1 + 0.5x_2 + 0.4x_3 \leq 2000$
             $3x_1 + 4x_2 + 3x_3 \leq 15000$
             $x_1 \geq 1000$
             $x_1, x_2, x_3 \geq 0$

   (ii)   The company should manufacture 1000 Amber chairs, 0 Beaton chairs and 4000 Countess chairs. The maximum total profit is then £54000.

   (iii) The slack variable associated with the wood constraint has an optimal value of 0 *i.e.* no wood is unused.

   (iv)   The slack variable associated with the fabric constraint has an optimal value of 100 *i.e.* 100m of the fabric is unused. Hence, the solution remains feasible if 100m of the fabric becomes damaged.

3. (i) Let z be the total number of Security Officers employed. Then, assigning the variables x_1 to x_6 as suggested in the question, the linear programming problem becomes :

Minimise $\quad z = x_1 + x_2 + x_3 + x_4 + x_5 + x_6$

Subject to : $\quad x_1 + x_6 \geq 4$

$x_1 + x_2 \geq 4$

$x_2 + x_3 \geq 6$

$x_3 + x_4 \geq 8$

$x_4 + x_5 \geq 6$

$x_5 + x_6 \geq 4$

$x_1, x_2, x_3, x_4, x_5, x_6 \geq 0$

(ii) The hotel should employ the Security Officers as follows :

Shift	1	2	3	4	5	6
Number of Security Officers	2	2	4	4	2	2

The minimum total number of Security Officers employed is then 16.

(iii) Let w be the total wage bill, C be the normal rate of pay per hour and $1.5C$ be the non-social hours rate per hour. The number of Security Officers working during shift 1 is $x_1 + x_6$. Hence, the linear programming model becomes :

Minimise $\quad w = 1.5C(x_1 + x_6) + C(x_2 + x_3 + x_4 + x_5)$

Subject to : $\quad x_1 + x_6 \geq 4$

$x_1 + x_2 \geq 4$

$x_2 + x_3 \geq 6$

$x_3 + x_4 \geq 8$

$x_4 + x_5 \geq 6$

$x_5 + x_6 \geq 4$

$x_1, x_2, x_3, x_4, x_5, x_6 \geq 0$

4. When a basic variable becomes non-basic its objective function coefficient remains zero or becomes positive (since zero or a positive value is added to its zero objective function coefficient during Jordan elimination). Hence, this variable will not be selected as the pivot variable at the next iteration *i.e.* this variable will not become basic again at the next iteration.

5. If a structural variable x_i is constrained to be non-positive then each occurrence of x_i must be replaced by a non-negative variable x_i' that is defined as :

$$x_i = -x_i'$$

The linear programming problem could then be solved in the usual way. The optimal value of x_i could be obtained from the optimal value of x_i'.

Exercises 3

1. (i) $z = 12$, $\begin{bmatrix} x_1 \\ x_2 \end{bmatrix} = \alpha \begin{bmatrix} 8 \\ 4 \end{bmatrix} + (1-\alpha) \begin{bmatrix} 4 \\ 8 \end{bmatrix}$ *i.e.* $\begin{bmatrix} x_1 \\ x_2 \end{bmatrix} = \begin{bmatrix} 4\alpha + 4 \\ 8 - 4\alpha \end{bmatrix}$ $\forall \alpha \in [0,1]$

   (ii) $z = -12$, $\begin{bmatrix} x_1 \\ x_2 \end{bmatrix} = \alpha \begin{bmatrix} 0 \\ 4 \end{bmatrix} + (1-\alpha) \begin{bmatrix} 12/5 \\ 14/5 \end{bmatrix}$ *i.e.* $\begin{bmatrix} x_1 \\ x_2 \end{bmatrix} = \begin{bmatrix} 12/5 - 12/5\alpha \\ 6/5\alpha + 14/5 \end{bmatrix}$ $\forall \alpha \in [0,1]$

   (iii) $z = -13$, $\begin{bmatrix} x_1 \\ x_2 \end{bmatrix} = \alpha \begin{bmatrix} 0 \\ 4 \end{bmatrix} + (1-\alpha) \begin{bmatrix} 8 \\ 0 \end{bmatrix}$ *i.e.* $\begin{bmatrix} x_1 \\ x_2 \end{bmatrix} = \begin{bmatrix} 8 - 8\alpha \\ 4\alpha \end{bmatrix}$ $\forall \alpha \in [0,1]$

2. Suppose that the optimal basic feasible solutions are $\underline{x}_1$ and $\underline{x}_2$. Then, the convex combination $\alpha \underline{x}_1 + (1-\alpha)\underline{x}_2$ is the equation of the optimal boundary of the feasible region.

   (i) Since the convex combination is part of the feasible region it must be a solution of the system of constraints .

   (ii) The convex combination satisfies the non-negativity conditions *i.e.* $\alpha \underline{x}_1 + (1-\alpha)\underline{x}_2 \geq 0$ $\forall \alpha \in [0,1]$. Hence, it must be a feasible solution.

   (iii) $\forall \alpha \in [0,1]$, the point $\begin{bmatrix} x_1 \\ x_2 \end{bmatrix}$ lies on the optimal boundary of the feasible region. Hence, the convex combination must be an optimal solution.

3. The linear programming problem is infeasible.

4. (i) The linear programming problem is unbounded.

   (ii) The linear programming problem is unbounded.

5. Question 1(i) : Not degenerate.

   Question 1(ii) : Degenerate.

   Question 1(iii) : Degenerate.

   Question 4(i) : Not degenerate.

   Question 4(ii) : Not degenerate.

6. (ii) The first tableau and the last tableau are the same *i.e.* the solution is cycling.

7. $x_1 = 1$, $x_2 = 0$, $x_3 = 1$, $x_4 = 0$, $z = -1$

Exercises 4a

1. $x_1 = 3$, $x_2 = 7/2$, $z = 5$

2. (i) The initial simplex tableau is :

$$\begin{bmatrix} 1 & 0 & 0 & -1 & -1 \\ 0 & 1 & 0 & -1 & 1 \\ 0 & 0 & 1 & 1 & 0 \end{bmatrix} \begin{bmatrix} -\overline{z} \\ x_3 \\ x_4 \\ x_1 \\ x_2 \end{bmatrix} = \begin{bmatrix} -3/2 \\ 1/2 \\ 3 \end{bmatrix}$$

(ii) The optimal transformation matrix is :

$$T = \begin{bmatrix} 1 & 1 & 2 \\ 0 & 1 & 1 \\ 0 & 0 & 1 \end{bmatrix}$$

Hence, the optimal simplex tableau is :

$$\begin{bmatrix} 1 & 1 & 2 \\ 0 & 1 & 1 \\ 0 & 0 & 1 \end{bmatrix} \begin{bmatrix} 1 & 0 & 0 & -1 & -1 \\ 0 & 1 & 0 & -1 & 1 \\ 0 & 0 & 1 & 1 & 0 \end{bmatrix} \begin{bmatrix} -\overline{z} \\ x_3 \\ x_4 \\ x_1 \\ x_2 \end{bmatrix} = \begin{bmatrix} 1 & 1 & 2 \\ 0 & 1 & 1 \\ 0 & 0 & 1 \end{bmatrix} \begin{bmatrix} -3/2 \\ 1/2 \\ 3 \end{bmatrix}$$

$$i.e. \quad \begin{bmatrix} 1 & 1 & 2 & 0 & 0 \\ 0 & 1 & 1 & 0 & 1 \\ 0 & 0 & 1 & 1 & 0 \end{bmatrix} \begin{bmatrix} -\overline{z} \\ x_3 \\ x_4 \\ x_1 \\ x_2 \end{bmatrix} = \begin{bmatrix} 5 \\ 7/2 \\ 3 \end{bmatrix}$$

3. Setting the non-basic variables to zero the linear programming problem reduces to :

$$\begin{bmatrix} 1 & \underline{c}_B & \underline{c}_N \\ 0 & B & N \end{bmatrix} \begin{bmatrix} -z \\ \underline{x}_B \\ \underline{0} \end{bmatrix} - \begin{bmatrix} -c_0 \\ \underline{b} \end{bmatrix}$$

$$i.e. \quad \begin{bmatrix} 1 & \underline{c}_B \\ 0 & B \end{bmatrix} \begin{bmatrix} -z \\ \underline{x}_B \end{bmatrix} = \begin{bmatrix} -c_0 \\ \underline{b} \end{bmatrix}$$

This system can be solved for the basic variables $\underline{x}_B$ by premultiplying each side by the inverse of the matrix of coefficients. Hence, the transformation matrix is the inverse of the matrix :

$$\begin{bmatrix} 1 & \underline{c}_B \\ 0 & B \end{bmatrix}$$

Let this inverse be the matrix be :

$$\begin{bmatrix} w & x \\ y & z \end{bmatrix}$$

Using the definition of the inverse matrix *i.e.* $MM^{-1} = I$, where I is the identity matrix :

$$\begin{bmatrix} 1 & \underline{c}_B \\ 0 & B \end{bmatrix}\begin{bmatrix} w & x \\ y & z \end{bmatrix} = \begin{bmatrix} 1 & 0 \\ 0 & 1 \end{bmatrix}$$

By multiplying out, equating the corresponding terms and solving for the components of the inverse matrix it can be shown that :

$$w = 1,\ x = -\underline{c}_B B^{-1},\ y = \underline{0}\ \text{and}\ z = B^{-1}$$

Substituting these, the inverse matrix *i.e.* the transformation matrix is :

$$T = \begin{bmatrix} 1 & -\underline{c}_B B^{-1} \\ \underline{0} & B^{-1} \end{bmatrix}$$

as required.

4. (i) Let x_1 be the amount of grade 1 ore purchased each year and x_2 be the amount of grade 2 ore purchased each year. Then, the linear programming model is :

Maximise $z = 500x_1 + 400x_2$

Subject to : $x_1 + x_2 \le 10000$

$x_1 \le 6000$

$x_2 \le 8000$

$0.3x_1 + 0.6x_2 \le 5000$

$0.5x_1 + 0.1x_2 \le 3200$

$x_1, x_2 \ge 0$

(ii) The foundry should buy 5500 tons of grade 1 ore and 4500 tons of grade 2 ore. The maximum profit is then £4,550,000.

Exercises 4b

1. (i) $x_1 = 45,\ x_2 = 0,\ z = 45$

(ii) $x_1 = 0,\ x_2 = 1,\ x_3 = 0,\ z = -5$

(iii) $x_1 = 2,\ x_2 = 0,\ z = 1$

2. The non-basic variable x_2 has a zero coefficient in the z-row of the optimal tableau.

3. Let the transformation matrix be the general matrix :

$$\begin{bmatrix} r & s & t \\ u & v & w \\ x & y & z \end{bmatrix}$$

The procedure used in question 3 in Exercises 4a produces the expression :

$$\begin{bmatrix} 1 & 0 & \underline{c}_B \\ 0 & 1 & \underline{d}_B \\ 0 & 0 & B \end{bmatrix}\begin{bmatrix} r & s & t \\ u & v & w \\ x & y & z \end{bmatrix} = \begin{bmatrix} 1 & 0 & 0 \\ 0 & 1 & 0 \\ 0 & 0 & 1 \end{bmatrix}$$

Multiplying out, equating the corresponding terms and solving for the components of the inverse matrix gives :

$$T = \begin{bmatrix} 1 & 0 & -\underline{c}_B B^{-1} \\ 0 & 1 & -\underline{d}_B B^{-1} \\ \underline{0} & \underline{0} & B^{-1} \end{bmatrix}$$

as required.

4. (i) Let z be the total yield, x_1 be the proportion of the fund invested in bonds, x_2 be the proportion of the fund invested in preference shares and x_3 be the proportion of the fund invested in ordinary shares. Then, the linear programming model is :

Maximise $\quad z = 6x_1 + 7x_2 + 12x_3$

Subject to : $\quad x_1 + 2x_2 + 4x_3 \leq 3$

$\qquad\qquad 6x_1 + 7x_2 + 12x_3 \geq 7$

$\qquad\qquad x_1 + x_2 + x_3 = 1$

$\qquad\qquad x_1, x_2, x_3 \geq 0$

(ii) The manager should invest 1/3 of the fund in bonds and 2/3 of the fund in ordinary shares *i.e.* she should not invest in preference shares. The maximum total yield is then 10%.

Exercises 5

1. Values $= \begin{bmatrix} 2 & 1 & 4 & 5 & 1 & 2 & 1 & 7 & 2 & 1 & 3 \end{bmatrix}$

Row Index $= \begin{bmatrix} 1 & 1 & 2 & 2 & 4 & 6 & 8 & 9 & 11 & 14 & 14 \end{bmatrix}$

Column Index $= \begin{bmatrix} 4 & 15 & 1 & 11 & 7 & 12 & 2 & 12 & 3 & 10 & 12 & 11 \end{bmatrix}$

2. (i)   A suitable scaled problem is :

Minimise $z = 4x_1 + 1.346x_2' - x_3'$

Subject to : $2x_1 + 5.614x_2' + 5x_3' \le 2746.01$

$-x_1 + 6.594x_2' + 4x_3' = 3037.40$

$7x_1 + 3x_2' + 6x_3' \le 1820.00$

$x_1, x_2', x_3' \ge 0$

This was obtained by multiplying through the third constraint by 1000, letting $x_2 = x_2'/1000$ and $x_3 = 1000x_3'$.

(ii)   The optimal solution of the scaled problem is : $x_1 = 0.0000$, $x_2' = 399.7489$, $x_3' = 100.3639$ and $z = 437.6981$.

(iii)   The optimal solution of the original problem is : $x_1 = 0.0000$, $x_2 = 0.3997489$, $x_3 = 100363.9000$ and $z = 437.6981$.

3. (i)   The initial tableau is :

-z	x_1	x_2	x_3	x_4	x_5	RHS
1.0000	0.9300	-2.0000	0.0000	0.0000	0.0000	0.0000
0.0000	0.9000	0.8500	-1.0000	0.0000	0.0000	2.3000
0.0000	-1.1000	0.9600	0.0000	-1.0000	0.0000	1.2000
0.0000	0.0000	0.9200	0.0000	0.0000	1.0000	3.0000

(ii)   Reading from top to bottom, the current basis is x_1, x_2 and x_3.

(iii)   The accurate tableau is :

-z	x_1	x_2	x_3	x_4	x_5	RHS
1.0000	0.0000	0.0000	0.0000	-0.8455	1.2917	4.8897
0.0000	1.0000	0.0000	0.0000	0.9091	0.9846	1.7549
0.0000	0.0000	1.0000	0.0000	0.0000	1.0870	3.2610
0.0000	0.0000	0.0000	1.0000	0.8182	1.7777	2.0513

4. (i)   The new T matrix is :

$$T = \begin{bmatrix} 1.0000 & 0.0000 & 0.8455 & 1.2917 \\ 0.0000 & 0.0000 & -0.9091 & 0.9486 \\ 0.0000 & 0.0000 & 0.0000 & 1.0870 \\ 0.0000 & -1.0000 & 0.8182 & 1.7777 \end{bmatrix}$$

(ii)   The matrix representation (with the basic variable x_5 listed first) is :

$$
\begin{bmatrix}
1.0000 & 0.0000 & 0.9300 & -2.0000 & 0.0000 & 0.0000 \\
0.0000 & 0.0000 & 0.9000 & 0.8500 & -1.0000 & 0.0000 \\
0.0000 & 0.0000 & -1.1000 & 0.9600 & 0.0000 & -1.0000 \\
0.0000 & 1.0000 & 0.0000 & 0.9200 & 0.0000 & 0.0000
\end{bmatrix}
\begin{bmatrix} -z \\ x_5 \\ x_1 \\ x_2 \\ x_3 \\ x_4 \end{bmatrix}
=
\begin{bmatrix} 0.0000 \\ 2.3000 \\ 1.2000 \\ 3.0000 \end{bmatrix}
$$

(iii) Multiplying each side of this expression by the T matrix from 4(i) gives :

$$
\begin{bmatrix}
1.0000 & 1.2917 & 0.0000 & 0.0000 & 0.0000 & -0.8455 \\
0.0000 & 0.9846 & 1.0000 & 0.0000 & 0.0000 & 0.9091 \\
0.0000 & 1.0870 & 0.0000 & 1.0000 & 0.0000 & 0.0000 \\
0.0000 & 1.7777 & 0.0000 & 0.0000 & 1.0000 & 0.8182
\end{bmatrix}
\begin{bmatrix} -z \\ x_5 \\ x_1 \\ x_2 \\ x_3 \\ x_4 \end{bmatrix}
=
\begin{bmatrix} 4.8897 \\ 1.7549 \\ 3.2610 \\ 2.0513 \end{bmatrix}
$$

i.e. the accurate tableau from question 3(iii).

Exercises 6

1. $x_1 = 0$, $x_2 = 0.5$, $z = -0.5$

2. $x_1 = -1$, $x_2 = 0$, $z = -2$

3.   Let z be the total estimated value of the components to the company, x_0 be a binary variable modelling the decision to either manufacture both component 1 and component 2 or not to manufacture either component 1 or component 2 and x_i be a binary variable modelling the decision to manufacture or not to manufacture component i. Then, the required linear programming model is :

Maximise $z = 7x_1 + 17x_2 + 11x_3 + 9x_4 + 21x_5$
Subject to : $3x_1 + 8x_2 + 5x_3 + 4x_4 + 10x_5 \leq 20$
$x_1 + x_2 - 2x_0 = 0$
$x_3 + x_4 \leq 1$
$x_0, x_1, x_2, x_3, x_4, x_5 \leq 1$
$x_0, x_1, x_2, x_3, x_4, x_5 \geq 0$
$x_0, x_1, x_2, x_3, x_4, x_5$ integer

4. (i) Using a penalty value of 1000 the linear programming problem can be written as :

Maximise $z = 2x_1 + 3x_2$

Subject to : $x_1 + x_2 \leq 20y_1 + 1000(1 - y_1)$ ---- (1)

$x_1 + x_2 \geq 20y_1$ ---- (2)

$x_1 - x_2 \leq 5y_2 + 1000(1 - y_2)$ ---- (3)

$2x_1 + 3x_2 \geq 25y_3$ ---- (4)

$y_1 + y_2 + y_3 \leq 2$ ---- (5)

y_1, y_2, y_3 binary, $x_1, x_2 \geq 0$

The optimal solution of this problem is $y_1 = 1$, $y_2 = 0$, $y_3 = 1$, $x_1 = 20$, $x_2 = 0$ and $z = 40$.

By setting $y_1 = 1$, $y_2 = 0$ and $y_3 = 1$ Solver has determined that the maximum value of z is obtained using constraints (1), (2) and (4).

(ii) The optimal solution in this case is $y_1 = 0$, $y_2 = 1$, $y_3 = 1$, $x_1 = 0$, $x_2 = 8.\dot{3}$ and $z = 25$.

Introducing the additional constraint $y_2 = 1$ has forced Solver to use constraints (3) and (4). It can be seen that the value of z obtained in this case is smaller than the value obtained in 4(i) *i.e.* that the solution obtained in 4(i) is correct.

5. (i) Using a penalty value of 1000 the linear programming problem can be written as :

Maximise $z = x_1 + x_2$

Subject to : $4x_1 + 5x_2 \leq 200y_1 + 1000(1 - y_1)$ ---- (1)

$x_1 + 4x_2 \geq 40y_2$ ---- (2)

$2x_1 + 3x_2 \leq 90y_3 + 1000(1 - y_3)$ ---- (3)

$2x_1 + 3x_2 \geq 90y_3$ ---- (4)

$y_1 + y_2 + y_3 \geq 2$ ---- (5)

y_1, y_2, y_3 binary, $x_1, x_2 \geq 0$

The optimal solution of this problem is $y_1 = 1$, $y_2 = 1$, $y_3 = 0$, $x_1 = 50$, $x_2 = 0$ and $z = 50$.

By setting $y_1 = 1$, $y_2 = 1$ and $y_3 = 0$ Solver has determined that the maximum value of z is obtained using constraints (1) and (2).

(ii) The optimal solution in this case is $y_1 = 1$, $y_2 = 1$, $y_3 = 1$, $x_1 = 45$, $x_2 = 0$ and $z = 45$.

Introducing the additional constraint $y_3 = 1$ has forced Solver to use all of the constraints. It can be seen that the value of z obtained in this case is smaller than the value obtained in 5(i) *i.e.* that the solution obtained in 5(i) is correct.

6. (i) Using a penalty value of 1000 the linear programming problem can be written as :

Maximise $z = 2x_1 + 3x_2$

Subject to : $2x_1 + x_2 \le 8$ ---- (1)

$x_1 + x_2 \le 6 + 1000y_1$ ---- (2)

$x_1 + 12x_2 \le 10 + 1000(1 - y_1)$ ---- (3)

y_1 binary, $x_1, x_2 \ge 0$

The optimal solution of this problem is $y_1 = 0$, $x_1 = 0$, $x_2 = 6$ and $z = 18$.

By setting $y_1 = 0$ Solver has determined that the maximum value of z is obtained using constraints (1) and (2).

(ii) The optimal solution in this case is $y_1 = 1$, $x_1 = 3.7391$, $x_2 = 0.5217$ and $z = 9.0435$

Introducing the additional constraint $y_1 = 1$ has forced Solver to use constraints (1) and (3). It can be seen that the value of z obtained in this case is smaller than the value obtained in 6(i) *i.e.* that the solution obtained in 6(i) is correct.

Exercises 7a

1. (i) Maximise $v = y_1 + 11y_2 + 14y_3$

Subject to : $2y_1 - y_2 + y_3 \le 1/2$

$-y_1 + 2y_2 + y_3 \le 1$

$y_1, y_2, y_3 \ge 0$

2. (i) Minimise $z = 1/2x_1 + x_2$

Subject to : $2x_1 - x_2 \ge 1$

$-x_1 + 2x_2 \ge 11$

$x_1 + x_2 \ge 14$

$x_1, x_2 \ge 0$

The dual of the dual is the primal (as stated in Theorem 1)

(ii) The dual is the easier problem to solve. It has fewer constraints and all of the constraints are type 1 so that it can be solved using a one-phase procedure.

(iii) $v = 67/6$, $y_1 = 0$, $y_2 = 1/6$, $y_3 = 2/3$.

(iv) The maximum value of v is the same as the minimum value of z.

3. (i) Maximise $v = y_1 + 3y_2 + 4y_3$

     Subject to : $2y_1 + 3y_2 - y_3 \leq 1$

                $-y_1 + y_2 + 2y_3 \leq 1$

                $-y_1 - y_2 + 2y_3 \leq -1$

                y_1 unconstrained, $y_2 \geq 0$, $y_3 \leq 0$

  (ii) Minimise $v = 1/2\, y_1 + 4y_2 + 5y_3 + 3y_4$

     Subject to : $y_1 + y_2 - y_3 + 2y_4 \leq 2$

                $y_1 + 2y_2 + 5y_3 + y_4 \geq -1$

                $y_1, y_3 \geq 0$, $y_2, y_4 \leq 0$

4. (i) The standard primal is :

     Minimise $z = x_1 + x_2 - x_3$

     Subject to : $2x_1 - x_2 - x_3 \geq 1$

                $-2x_1 + x_2 + x_3 \geq -1$

                $3x_1 + x_2 - x_3 \geq 3$

                $x_1 - 2x_2 - 2x_3 \geq -4$

                $x_1, x_2, x_3 \geq 0$

     To reproduce the answer given for question 3(i), find the dual and then let $y_1 - y_2 = t_1$, $y_3 = t_2$ and $y_4 = -t_3$.

  (ii) Let $x_1' = -x_1$. Then, the standard primal is :

     Minimise $\bar{z} = 2x_1' + x_2$

     Subject to : $x_1' - x_2 \geq -1/2$

                $-x_1' + 2x_2 \geq 4$

                $-x_1' - 5x_2 \geq -5$

                $-2x_1' + x_2 \geq 3$

                $x_1', x_2 \geq 0$

     To reproduce the answer given for question 3(ii), find the dual and then let $y_2' = -y_2$ and $y_4' = -y_4$.

5. The pivot selection rule breaks down *i.e.* the dual problem is unbounded. Hence, by Theorem 2 the primal problem is infeasible.

6. $z = -106$, $x_1 = 0$, $x_2 = -13/2$, $x_3 = -7/2$.

7. $z = -12$, $x_1 = 0$, $x_2 = -2$, $x_3 = 0$.

8. $z = 5$, $x_1 = -1$, $x_2 = 2$.

Exercises 7b

1. (i) $x_1 = 0$, $x_2 = 2$, $x_3 = 1$, $z = 19$

   (ii) $x_1 = 35/3$, $x_2 = 55/3$, $x_3 = 25/3$, $z = -30$. This solution is non-unique.

2. (ii) The only feasible solution is the optimal solution.

3. (i) Let z be the total cost of the scrap purchased, x_1 be the number of tonnes of Grade 1 purchased, x_2 be the number of tonnes of Grade 2 purchased and x_3 be the number of tonnes of Grade 3 purchased. Then, the linear programming model becomes :

   Minimise $\quad z = 50x_1 + 60x_2 + 40x_3$
   Subject to : $\quad 0.3x_1 + 0.4x_2 + 0.2x_3 = 3000$
   $\quad\quad\quad\quad 0.2x_1 + 0.3x_2 + 0.4x_3 = 2500$
   $\quad\quad\quad\quad x_1, x_2, x_3 \geq 0$

   (ii) $x_1 = 0$, $x_2 = 7000$, $x_3 = 1000$, $z = 460000$ *i.e.* the company should purchase 0 tonnes of Grade 1, 7000 tonnes of Grade 2 and 1000 tonnes of Grade 3. The total cost is then £460,000.

4. (i) $x_1 = 0$, $x_2 = 0.2$, $x_3 = 0.3$, $z = 0.8$

   (ii) The pivot selection rule breaks down *i.e.* the problem is infeasible.

   (iii) $x_s = 0$ and $z^* = f(M)$ *i.e.* the problem is unbounded.

   (iv) In the optimal tableau a non-basic variable has a zero coefficient in the z-row *i.e.* the solution is non-unique. The first optimal solution is $x_1 = 8$, $x_2 = 4$ and $z = 12$. Choosing the non-basic variable as the pivot variable, the pivot row using the pivot selection rule used with the simplex method and then performing Jordan elimination, the second optimal solution is $x_1 = 4$, $x_2 = 8$ and $z = 12$. The general solution of this linear programming problem can be written as a convex combination of these two optimal solutions *i.e.* as :

   $$\alpha \underline{x}_1 + (1-\alpha)\underline{x}_2$$

   where :

   $$\underline{x}_1 = \begin{bmatrix} 8 \\ 4 \end{bmatrix} \text{ and } \underline{x}_2 = \begin{bmatrix} 4 \\ 8 \end{bmatrix}$$

   Hence, the general solution of the linear programming problem is :

$$i.e. \quad \alpha \begin{bmatrix} 8 \\ 4 \end{bmatrix} + (1-\alpha) \begin{bmatrix} 4 \\ 8 \end{bmatrix} \quad i.e. \quad \begin{bmatrix} 4\alpha+4 \\ 8-4\alpha \end{bmatrix} \quad \forall \alpha \in [0,1]$$

(v)   The pivot selection rule breaks down *i.e.* the problem is infeasible.

5. (i)   Here $n=2$ and $m=5$ *i.e.* $n<4m$. Hence, the simplex method should be used.

(ii)   The initial solution of this problem is neither optimal nor feasible. Hence, the artificial constraint technique must be used.

(iii) Both constraints are type 3. Hence, the dual simplex method should be used.

(iv) Here $n=9$ and $m=2$ *i.e.* $n>4m$. Hence, the revised simplex method should be used.

Exercises 8

1. $z=10$, $x_1=7$, $x_2=1$

2. $z=66100$, $x_1=122$, $x_2=78$

3. (i) The possible cutting patterns and trim wastes are :

Cutting Pattern	1	2
Number of **25-inch** Rolls	1	0
Number of **21-inch** Rolls	1	2
Trim Waste (inches)	2	6

Let w be the total trim waste, x_1 be the number of large rolls cut according to cutting pattern 1 and x_2 be the number of large rolls cut according to cutting pattern 2. Then, the required integer linear programming model becomes :

$$\text{Minimise} \quad w = 2x_1 + 6x_2$$
$$\text{Subject to :} \quad x_1 \geq 20$$
$$x_1 + 2x_2 \geq 50$$
$$x_1, x_2 \geq 0$$
$$w, x_1, x_2 \text{ integer}$$

(ii) $w=100$, $x_1=50$, $x_2=0$ *i.e.* the paper should cut 50 large rolls according to cutting pattern 1 and 0 large rolls according to cutting pattern 2. This will produce a minimum trim waste of 100 inches.

4. $z=11$, $x_1=2$, $x_2=3$ <u>or</u> $z=11$, $x_1=5$, $x_2=2$

5. $z=10$, $x_1=2$, $x_2=2$

A. M. FITZHARRIS

6. (i) Let w be the total trim waste, x_1 be the number of large rolls cut according to cutting pattern 1 and x_2 be the number of large rolls cut according to cutting pattern 2, *etc.* Then, the required integer linear programming model becomes :

$$\text{Minimise} \quad w = 23x_1 + 27x_2 + 17x_3 + 31x_4 + 21x_5 + 11x_6 + 25x_8 + 15x_9 + 5x_{10}$$
$$\text{Subject to :} \quad 3x_1 + 2x_2 + 2x_3 + x_4 + x_5 + x_6 \geq 180$$
$$x_2 + 2x_4 + x_5 + 3x_7 + 2x_8 + x_9 \geq 90$$
$$2x_3 + 2x_5 + 4x_6 + x_7 + 2x_8 + 4x_9 + 6x_{10} \geq 90$$
$$x_1, x_2, x_3, x_4, x_5, x_6, x_7, x_8, x_9, x_{10} \geq 0$$
$$w, x_1, x_2, x_3, x_4, x_5, x_6, x_7, x_8, x_9, x_{10} \text{ integer}$$

(ii) $w = 1380$, $x_1 = 60$, $x_7 = 90$ *i.e.* the paper should cut 60 large rolls according to cutting pattern 1 and 90 large rolls according to cutting pattern 7. The other cutting patterns are not used. This will produce a minimum trim waste of $1380cm$.
This solution is <u>not</u> a sensible because the additional 180 $60cm$ rolls are not required *i.e.* are trim waste. A better solution can be obtained by replacing the "$\geq$" operators in the constraints with "$=$" signs.

Exercises 9

1. $B^{*-1} = \begin{bmatrix} 1 & 1 & 1 \\ 0 & 2 & -1 \\ 0 & -1 & 1 \end{bmatrix}$

2. $B^{*-1} = \begin{bmatrix} 1 & 3 & 0 \\ -1 & -2 & 0 \\ 1 & 1 & 1 \end{bmatrix}$

3. (i) $x_1 = 3/4$, $x_2 = 1/4$, $x_3 = 3/4$, $x_4 = 1/4$, $x_5 = 0$, $z = 5/4$

   (ii) b_1 can take any value in the range $[1/4, \infty)$ without affecting the optimal basis.

   b_2 can take any value in the range $[3/4, \infty)$ without affecting the optimal basis.

   b_3 cannot be varied without producing a change in the optimal basis.

   b_4 can take any value in the range $[1/2, 3/2]$ without affecting the optimal basis.

   (iii) (a) $z^+ = 5/4$
   (b) $z^+ = 5/4$
   (c) b_3 cannot be varied without producing a change in the optimal basis.
   (d) $z^+ = 9/8$

271

4. (i) $T = \begin{bmatrix} 1 & 0 & 0 & 0 & 1/2 \\ 0 & 1 & 0 & 1 & 1 \\ 0 & 0 & 1 & 0 & -2 \\ 0 & 0 & 0 & 0 & 1/2 \\ 0 & 0 & 0 & -1 & 1/2 \end{bmatrix}$

(ii) $x_1 = 50$, $x_2 = 0$, $z = 50$

5. $x_1 = 0$, $x_2 = 40$, $z = 200$

6. $x_1 = 3$, $x_2 = 7/2$, $z = 43/2$

7. $-5/2 \le c_2 \le -1$

8. $x_1 \approx 323.33$, $x_2 \approx 21.67$, $z \approx 134.75$

9. $x_1 = 252.0000$, $x_2 = 210.0000$, $z = 16800.0000$

Exercises 10a

(i)   See answer to Exercises 2, Question 3(ii)

(ii)   See answer to Exercises 6, Question 4(i)

(iii) See answer to Exercises 6, Question 5(i)

(iv)   See answer to Exercises 6, Question 6(i)

(v)   See answer to Exercises 8, Question 6(ii)

Exercises 10b

3.

	A	B	C	D	E	F	G	H	I	J
1	**Mix Information**									
2										
3	**Material Number**	**Proportion**	**14.00mm**	**10.00mm**	**6.30mm**	**3.35mm**	**2.36mm**	**425mic**	**150mic**	**75mic**
4	2	30.54%	30.54	22.08	1.04	0.00	0.00	0.00	0.00	0.00
5	3	0.00%	0.00	0.00	0.00	0.00	0.00	0.00	0.00	0.00
6	4	69.46%	69.46	69.46	69.46	68.00	43.76	24.10	9.03	2.78
7	5	0.00%	0.00	0.00	0.00	0.00	0.00	0.00	0.00	0.00
8	6	0.00%	0.00	0.00	0.00	0.00	0.00	0.00	0.00	0.00
9	7	0.00%	0.00	0.00	0.00	0.00	0.00	0.00	0.00	0.00
10										
11	**Total % :**	100.00	100.00	91.54	70.50	68.00	43.76	24.10	9.03	2.78
12										
13										
14		**Min %**	100	82	69	51	32	15	7	2
15		**Max %**	100	100	83	68	48	29	11	5
16										
17		**Cost of Mix :**	£5.49	**Per Tonne (Excluding plant mixing cost)**						

4.

	A	B	C	D	E	F	G	H
1	**Summary Report**							
2								
3		**Customer :**	Silverstone Motor Racing Circuit					
4	**Material Description :**		Special 10mm Asphalt Wearing Course					
5		**Date :**	19th February 2021					
6		**Cost of Mix :**	£5.49					
7	**Total Tonnes Required :**		5,000					
8								
9	**Material Number**	**Description**	**Supplier**	**Price Per Tonne**	**Proportion**	**Kg Per Tonne**	**Tonnes Required**	**Cost**
10	2	10mm Granite	Roadstone Aggregates Ltd.	£8.09	30.54%	305.41	1527.07	£12,353.98
11	3	6mm Granite	Roadstone Aggregates Ltd.	£8.47	0.00%	0.00	0.00	£0.00
12	4	Course Sand	Hall Aggregates	£4.35	69.46%	694.59	3472.93	£15,107.25
13	5	Medium Sand	Beds Silica	£4.63	0.00%	0.00	0.00	£0.00
14	6	Fine Sand	Kendall & Prestwick	£4.94	0.00%	0.00	0.00	£0.00
15	7	Limestone Filler	Stevenson Minerals	£10.60	0.00%	0.00	0.00	£0.00
16								
17							**Total :**	£27,461.24

Exercises 11

1. The optimal assignment is :

Worker	Task
1	E
2	C
3	B
4	D
5	A

The total time taken is 21 hours.

2. The optimal assignment is :

Truck	Building Site
I	E
II	A
III	D
IV	C
V	B

The total transportation cost is £210.

3. The optimal assignment is :

Taxi Location	Customer Location
A	3
C	1
D	2

The total distance travelled is 7 miles.

4. The optimal assignment is :

Operator	Machine
Dave	Drill
Sue	Milling Machine
Ben	Lathe
Sandeep	Grinder

The total cost is £14/hour.

5. The optimal sales strategy is :

Garment	Country
Hoodie	Hong Kong
Sweatshirt	Indonesia
T-Shirt	Taiwan
Shorts	Cambodia
Cap	Vietnam
Socks	Malaysia

The total profit is £72.

6. The optimal assignment is :

Employee	Office
Yvonne Percival	219
James Kingsley	211
Emily Deanus	201
Graham Roberts	231
Derek Johnson	212

Exercises 12a

1.

Factory	Warehouse	Number of Steel Panels
1	1	10
1	3	5
2	1	15
2	2	15
3	3	15
Total shipping cost : £460		

2.

Refinery	Service Station	Number of 1000 Litre Units
1	Sainsbury's	500
1	Tesco	100
2	ASDA	250
2	Tesco	150
3	Tesco	500
Total shipping cost : £8800		

The optimal shipping pattern is non-unique.

3.

Grade	Valve	Number of Craftsmen
Junior	Type 3	5
Senior	Type 2	10
Principal	Type 1	10
Principal	Type 2	5
Principal	Type 3	15
Total manufacturing cost : £1835		

4.

Contract	Pay Station	Number of Employees
C	1	100
A	2	5
B	2	218
A	3	161
C	3	49
D	3	190
Total time lost : 2452 minutes		

The optimal shipping pattern is non-unique.

Exercises 12b

1.

Factory	Warehouse	Number of Steel Panels
1	3	15
2	1	15
2	2	15
3	1	10
3	3	5

2.

Refinery	Service Station	Number of 1000 Litre Units
1	Sainsbury's	350
1	Tesco	250
2	Sainsbury's	150
2	ASDA	250
3	Tesco	500
Total shipping cost : £8800		

The optimal shipping pattern is non-unique.

3.

School	Area	Number of Pupils
Barnwell	Bedwell	150
Shephalbury	Chells	250
Heathcote	Bedwell	250
Total time taken : 4000 minutes		

4.

School	Area	Number of Pupils
Barnwell	Bedwell	150
Shephalbury	Chells	250
Heathcote	Bedwell	250
Total time taken : 4000 minutes		

5. (a) Minimise
$$z = \sum_{i=1}^{n}\sum_{j=1}^{m} c_{ij}x_{ij}$$

   Subject to :
$$\sum_{j=1}^{m} x_{ij} \le s_i, \quad i=1,\dots,n \quad \textit{i.e. supplies must not be exceeded.}$$

$$\sum_{i=1}^{n} x_{ij} = d_j, \quad j=1,\dots,m \quad \textit{i.e. demands must be met.}$$

$$x_{ij} \ge 0 \quad \forall i,j$$

   (b) Minimise
$$z = \sum_{i=1}^{n}\sum_{j=1}^{m} c_{ij}x_{ij}$$

   Subject to :
$$\sum_{j=1}^{m} x_{ij} = s_i, \quad i=1,\dots,n \quad \textit{i.e. supplies must be exhausted.}$$

$$\sum_{i=1}^{n} x_{ij} \le d_j, \quad j=1,\dots,m \quad \textit{i.e. demands must not be exceeded.}$$

$$x_{ij} \ge 0 \quad \forall i,j$$

6.

Mine	Power Station	Number of Coal Wagons
Barnsley	Glasgow	14
Harrogate	Falkirk	12
Harrogate	Aberdeen	6
Doncaster	Paisley	26
Doncaster	Aberdeen	6
Total shipping cost : £626		

7. (a)

Factory	Warehouse	Number of Sets
1	1	5
1	2	15
2	1	5
Total shipping cost : 2250*p i.e.* £22.50		

(c) The totals at factory 2 and warehouse 2 can be increased by 5 sets. The new optimal shipping pattern is :

Factory	Warehouse	Number of Steel Panels
1	2	20
2	1	10
Total shipping cost : 400*p i.e.* £4.00		

The resulting cost saving is 2250*p* – 400*p* = 1850*p i.e.* £18.50

8.

Warehouse	Supermarket	Number of Boxes
Acton	Bournemouth	200
Acton	Brighton	800
Fulham	Bournemouth	700
Wapping	Portsmouth	500
Wapping	Southampton	400
Total shipping cost : £15500		

9.

Factory	Warehouse	Number of Steel Panels
1	3	15
2	1	15
2	2	15
2	2	15
3	1	10
3	3	5
Total shipping cost : £470		

10.

From	To	Number of Cars
Los Angeles	Denver	1000
Detroit	Denver	1300
Detroit	Miami	200
New Orleans	Miami	1200
Total shipping cost : $313200		

11.

Factory	Warehouse	Number of Steel Panels
1	1	5
1	3	10
2	1	15
2	2	15
3	1	5
Total shipping cost : £465		

References

Alevras, D. and Padbury, M., 2001. *Linear Optimisation and Extensions : Problems and Solutions.* New York. Springer.

Bazaraa, M.S., Jarvis, J.J. and Sherali, H.D., 1990. *Linear Programming and Newtork Flows* (2nd Ed.). New York. Wiley.

Beasley, J.E. (Ed.)., 1996. *Advances in Linear and Integer Programming.* Oxford. Oxford Science.

Dantzig, G.B. and Mukund, M.N., 1997. *Linear Programming 1 : Introduction.* New York. Springer.

Dantzig, G.B. and Mukund, M.N., 1997. *Linear Programming 2 : Theory and Extensions.* New York. Springer.

Evar, D. and Tucker, A.W., 1993. *Linear Programs and Related Problems* (2nd Ed). London. Academic Press.

Gartner, B. and Matousek, J., 2007. *Understanding and Using Linear Programming.* New York. Springer.

Gass, S.I., 1990. *An Illustrated Guide to Linear Programming.* New York. Dover Publications Inc.

Gass, S.I., 2003. *Linear Programming : Methods and Applications* (5th Ed.). New York. Dover Publications Inc.

Karloff, H., 2008. *Linear Programming.* Basel. Birkhauser.

Kolman, B. and Beck, R.E., 1995. *Elementary Linear Programming with Applications* (2nd Ed.). London. Academic Press.

Murty, K.G., 1983. *Linear Programming.* New York. Wiley.

Padbury, M., 1999. *Linear Optimisation and Extensions* (2nd Ed.). New York. Springer.

Schrijver, A., 1998. *Theory of Linear and Integer Programming.* New York. Wiley.

Sultan, A., 1993. *Linear Programming : An Introduction with Applications* (2nd Ed.). London. Academic Press.

Taha, H.A., 2017. *Operations Research : An Introduction* (10th Ed.). London. Pearson.

Vajda, S., 1981. *Linear Programming : Algorithms and Applications.* New York. Springer.

Vanderbei, R.J., 2014. *Linear Programming : Foundations and Extensions* (4th Ed.). New York. Springer.

Vaserstein, L.N., 2002. *Introduction to Linear Programming.* London. Pearson Education.

Williams, H.P., 2013. *Model Building in Mathematical Programming* (5th Ed.). New York. Wiley.

Williams, H.P., 1993. *Model Solving in Mathematical Programming*. New York. Wiley.

Other Sources
Some of the examples presented in this book are taken from notes I was given as an undergraduate student in the early 1980s. Unfortunately, after nearly 40 years it is not possible to identify the source of these examples. However, should the authors wish to receive due credit they are invited to contact the author via the publisher so that appropriate references can be included in a future edition of this book.

Index

www.ingramcontent.com/pod-product-compliance
Lightning Source LLC
Chambersburg PA
CBHW050105220326
41598CB00043B/7384